Consulting Editor

George A. Anastassiou
Department of Mathematical Sciences
The University of Memphis

Alexander Vollert

A Stochastic Control Framework for Real Options in Strategic Evaluation

Birkhäuser
Boston • Basel • Berlin

Alexander Vollert
Universität Karlsruhe (TH)
Postfach 69 80
Karlsruhe, D-76128
Germany

and

McKinsey & Company, Inc.
Birkenwaldstr. 149
70191 Stuttgart
Germany

Library of Congress Cataloging-in-Publication Data

Vollert, Alexander.
    A stochastic control framework for real options in strategic valuation / Alexander Vollert.
      p. cm.
    Includes bibliographical references and index.
    ISBN 0-8176-4258-7 (alk. paper) – ISBN 3-7643-4258-7 (alk. paper)
      1. Real options (Finance)–Mathematical models. 2. Corporations–Valuation. 3.
Business enterprises–Valuation. 4. Capital investments–Decision making–Simulation
methods. I. Title.

    HG4028.V3 V652 2003
    332.63–dc21
                                                2002033288

AMS Subject Classifications: 60Gxx

Printed on acid-free paper.
©2003 Birkhäuser Boston

Birkhäuser

ISBN 0-8176-4258-7   SPIN 10844820
ISBN 3-7643-4258-7

Reformatted from author's files by TEXniques, Inc.
Printed in the United States of America.

9 8 7 6 5 4 3 2 1

Birkhäuser Boston • Basel • Berlin
A member of BertelsmannSpringer Science+Business Media GmbH

Contents

List of Figures

List of Tables

Acknowledgment

I wish to extend my deep gratitute to those who influenced me a great deal and helped me prepare this text.

My first acknowledgement is a deep one. Prof. Dr. Svetlozar T. Rachev (University of Karlsruhe, Germany, and University of California, Santa Barbara) was always generous with his time and counsel; his help and interest are acknowledged with grateful thanks. Along the way, my thinking was influenced by many stimulating discussions with Prof. Dr. Karl-Heinz Waldmann and Prof. Dr. Kuno Egle (both University of Karlsruhe). At various stages their role as a sounding board for various pieces proved extremely fruitful and is deeply acknowledged.

My former colleagues at University of Karlsruhe deserve special mention for their intellectual guidance. Dr. Elke Sennewald and Dr. Christian Peter were always inspirational discussion partners; we had a great time together in academia. My current employer, McKinsey & Company, Inc., deserves acknowledgement for providing me with the opportunity to finish this work.

I am grateful to Prof. Dr. George Anastassiou (University of Memphis) for recommending this text for publication. Furthermore, I would like to thank the Executive Editor, Ann Kostant of Birkhäuser, who worked with me on this book.

As expected, my greatest gratitude is to my family. Without the constant support of my parents throughout my life this book would not have seen the light of day. For her enduring guidance, encouragement, love and support, I would especially like to thank my wife Christina. Most important is my daughter Johanna Deborah Vollert, who stands by me enduring late nights and weekends, makes it all worthwhile and ensures that my life is full of love and excitement.

Finally, I would like to dedicate this book to my late father, Ulrich Vollert.

*A Stochastic Control Framework
for Real Options
in Strategic Evaluation*

Chapter 1

Overview

Taking an option-based approach is not simply a matter of using a new set of valuation equations and models. It requires a new way of framing strategic decisions.

—Martha Amram and Nalin Kulatilaka

1.1 Background and Objectives of the Study

An option is, in essence, the right to choose whether or not to take an action now or at some time in the future.[1] Broadly speaking, options can be divided into two types: financial options and real options. Financial options are rights to buy, sell, or exchange claims on traded securities, such as stocks and bonds. The Nobel prize winning works of Black/Scholes [27] and Merton [165] have been the basis for an extensive literature in finance, analyzing a growing variety of options and claims on financial securities.

Myers [176] first suggested that the market value of a firm consists of the present value of its expected cash flows and the value of the firm's growth opportunities. Because of the analogy with financial options, Myers described the firm's growth opportunities as the firm's real options. The theory of real options provides a tool to analyze managerial flexibility in the presence of uncertainty about the future in giving up investment expenditures in return for project value. Since the costs of the investment are at least partially sunk, the investment is irreversible. Managerial flexibility analyzed as real options can be an additional source of value for a firm facing uncertainty about future developments and irreversibility of its investment expenditures.

In 1985 the theoretical foundations for the real option literature were published. Brennan/Schwartz [34] and McDonald/Siegel [159] valued risky invest-

---

[1] See Sanchez [196].

ment projects where there is an option to temporarily and costlessly shut down production. Moreover, Brennan/Schwartz showed the practical merits of real options by an application to the value of a copper mine. They value the option to mothball the copper mine when prices are low and to later reactivate the mine if prices rise. Subsequently, McDonald/Siegel [160] developed a model that forms the basis for many current applications of real option theory. They showed that in an uncertain market, a firm should postpone an irreversible investment until the value of the investment exceeds the costs of the investment by the option value of waiting to invest. They derive a critical threshold at which it is optimal to invest. Since the option value increases with uncertainty, the critical value and hence the timing of investment increases with uncertainty as well.

In the late 1980s and 1990s, the theory of real options rapidly expanded and has developed into an actively growing field of research. Originally on the natural resources industry focused recent publications in this field deal with applications in manufacturing, R&D, electric utility industry, regulated firms, M&A, leases, international investments, real estate and land development as well as corporate strategy, to name a few.[2] From a valuation perspective several generic real options have been identified, like the option to abandon a project, the option to stage investments, the option to switch inputs or outputs, the option to expand or contract the scale of a project, and strategic growth options.[3] Furthermore, various limitations of the financial option analogy have been identified which led, for example, to the integration of game theory into real option analysis in order to model the influence of competition on real option values.[4]

Although the publication of several books on real options[5] and articles in management magazines[6] indicate that the real options idea is also about to enter the world of corporate practitioners, there still exist some serious obstacles constraining management from widely applying real option analysis in practice. One of the most serious obstacles is that corporate managers, practitioners, and even many academics do not have the required mathematical skills to use the models comfortably. Furthermore, many of the required modelling assumptions are often and consistently violated in practical real option applications. Finally, the scope of applications is limited by the simplified modelling assumptions often used in academic research. Especially, the stylized examples with few incorporated real options that can be found in the literature are not adequate for real life applications.

Consequently, there is a need for more practically oriented decision making

---

[2]See Lander/Pinches [139], p. 539, for an overview of recent applications that have been dealt with in the literature.

[3]See Trigeorgis [225], p. 2f.

[4]See the discussion in Section 2.2.2.

[5]See the fundamental books of Dixit/Pindyck [72], Trigeorgis [225] and Amram/Kulatilaka [7] on real option analysis, the article collections edited by Trigeorgis [227] and Brennan/Trigeorgis [35] as well as the various special issues of scientific magazines on real options.

[6]See especially the articles by Dixit/Pindyck [73], Luehrman [147, 148], and Amram/Kulatilaka [6] in *Harvard Business Review* as well as Leslie/Michaels [143] and Copeland/ Keenan [59, 58] in *The McKinsey Quarterly*.

frameworks which:

- can model and value real world investment opportunities having many time periods, multiple interacting real options, multiple uncertainties, state and/or time dependent payouts etc.,

- are easier to understand and implement and therefore more likely to be used,

- can make real option valuation available to financial practitioners, corporate managers, and other decision-makers, and

- focus on determining optimal decision strategies rather than obtaining an exact valuation.

Therefore, the purpose of this book is to develop both a stochastic control framework that is capable of dealing with complex real option interactions and a graphical decomposition method for real option models which can be easily communicated to corporate practitioners and subsequently be transformed into numerical or analytical solution procedures.

The resulting objectives of the book are:

- to emphasize the importance of flexibility, uncertainty, and irreversibility in investment decisions,

- to show that traditional capital budgeting methods are not appropriate when the decision maker faces uncertainty and irreversibility,

- to present a new framework for real options which combines a management and a valuation perspective and which extends the traditional definitions of real options that can be found in the literature,

- to introduce and develop stochastic control methods as a means to evaluate complex decision problems with different interacting real options,

- to display complex real option models as graphical representations of their contingency structure,

- to develop a procedure for transforming graphical representation into numerical solution schemes,

- to integrate recent advances in the real option literature into the proposed stochastic control framework, and

- to give several examples of how to work within the stochastic control framework.

1.2   Organization of the Study

The book is organized as follows. Chapter 2 gives an introduction to the ideas of real option analysis. After exploring why managerial flexibility adds value to the firm in the presence of uncertainty and irreversibility, the lack of standard capital budgeting tools in taking account of uncertainty and irreversibility is discussed. The expanded net present value as a means to reflect the value of managerial flexibility is introduced and the general assumptions for valuing real options with financial option pricing theory are presented. Moreover, a new framework for classifying real options from a management as well as a valuation perspective is developed. Finally, the advantages and disadvantages of the real option approach are discussed in detail already motivating the creation of models that will be capable of dealing with complex real option interactions.

Chapter 3 presents the stochastic control framework for pricing multiple real options. It starts with the nature of real option interactions and a review of the work that has been done in the field. It is claimed that virtually every investment decision model with different embedded real options can be decomposed into combinations of generalized switching options and generalized timing options if a continuous time framework is used.

The chapter also presents mathematical formulation of the corresponding impulse control and optimal stopping problems represented by the generalized switching and generalized timing option, respectively, and develops verification theorems to ensure the optimality of the results. As it turns out the solution to real option interaction models boils down to solving systems of quasi-variational inequalities, which usually needs to be done numerically.

In Chapter 4 it is shown that the market value of a generalized switching option is, in fact, the solution to an equivalent impulse control problem which establishes close resemblance between modern option pricing theory and stochastic control, if the link to capital markets is properly defined. Since the mathematical machinery of stochastic control theory is not suited for communication to corporate practitioners, Section 4.2 provides a simple procedure for decomposing interaction models into graphical representation schemes — the contingency structure of option interactions — that can easily be transformed into numerical solution procedures. A stylized example for transforming a generalized timing option into the graphical representation and solving it analytically is presented in Section 4.3.

Chapter 5 deals with the problem of integrating recent advances in the theory of real options into the stochastic control framework. Especially, models for endogenous and exogenous competition are included in the analysis. For the case of endogenous competition a new option game model is developed that allows for determining optimal entry and exit strategies in a duopolistic setting if, in addition, the capacity choice by both players is considered. Another extension that has been recently discussed in the real option literature is the consideration of time lags between the exercise date of the real option and the date when the decision finally takes effect. Section 5.2 shows how such a situation can be handled in the stochastic control framework.

In order to demonstrate the use of the stochastic control framework in a more realistic decision problem, Chapter 6 presents a case study of flexible capacity in the manufacturing industry. It is assumed that the firm is exposed to a stochastic product life cycle and has several combinations of real options in order to adjust its operating strategy to its uncertain environment. Starting with a general discussion of the importance of real options in operations, the corresponding mathematical models as well as the graphical representations for different combinations of real options are derived in Section 6.2. Afterwards a numerical solution procedure resting on finite differences is given, which allows for solving the considered real option interaction model numerically. Section 6.4 displays numerical solutions for different manufacturing models along with optimal operating strategies of the firm, given as thresholds for taking action. A sensitivity analysis with respect to the main input parameters is performed. In order to gain a deeper understanding of the different incorporated real options, a simulation study is carried out in Section 6.5 which displays the optimal operating strategies along certain trajectories. Moreover, the simulation study provides further insights into the nature of impulse control problems. Afterwards stochastic dominance criteria are employed to identify superior operating strategies resulting from different combinations of real options.

Finally, Chapter 7 summarizes the main results of the book, draws some general conclusions, and points out several directions for further research.

Chapter 2

Introduction to Real Options

2.1 Basic Idea

This section gives an overview of the basic ideas of the *real option approach* to capital budgeting in firms. It deals with the importance of recognizing flexibility in capital investments, and how traditional ways of valuing investments have largely neglected the benefits of flexibility in the presence of uncertainty and irreversibility. Flexibility in investment decision making can limit the downside potential of an investment while exploiting its upside potential. In its asymmetric structure, it is shown that flexibility possesses option-like features that allow for interpreting managerial flexibility as options on real assets which can be valued using option pricing methods well known in financial markets. The resulting additional value introduced by flexibility can be operationalized by the notion of *expanded net present value*, which covers the sources of value coming from the (static) project without flexibility and the value of all relevant managerial real options.

2.1.1 Why Flexibility Adds Value

Considering Flexibility, Uncertainty, and Irreversibility

There are only two situations where managerial flexibility[1] does not affect investment decisions of firms. The first is if the decision maker has *perfect information*. If all relevant information is known with certainty, the decision process is reduced to choosing the best among all available alternatives. There is no need for active management in order to improve the performance of the investment. As a consequence, the optimal path of all future decisions is already predetermined. Certainly this is not the case in the business environment management faces at present.

---

[1] See Busby/Pitts [37], p. 1, for an introduction to the term managerial flexibility.

The second is if all decisions that have been made up to the present are *completely reversible*. A decision that can be reversed at any time without any regret or payment can be made as if no risk were present.

Since flexibility of management is in both cases worthless, and investment decisions can be made without any risk, capital budgeting decisions simply reduce to discounting future cash flows with the riskless interest rate.

From that point of view classical capital budgeting and valuation techniques like the discounted cash flow approach implicitly assume that the decision maker faces uncertainty as well as irreversibility. But since flexibility is especially important in the presence of uncertainty and irreversibility, for any valuation technique that is used as a quantification tool in the decision making process, it is a necessity to consider flexibility *explicitly*. In extreme situations, the incorporation of flexibility whenever quantitative capital budgeting tools are employed is a *conditio sine qua non*.

Therefore, the meaning of flexibility for investment decisions is predetermined whenever the decision has to be made under uncertainty and irreversibility. Uncertainty exists when the decision maker does not have perfect information about both the current state he is in and the consequences his decision will have contingent on the current state.

Uncertainty is often distinguished between situations where the decision maker is able to specify a probability distribution of the uncertain state[2] and situations where such a probability distribution is unknown. The latter case is hard to tackle from a scientific point of view, since it is simply impossible to build decision models without being able to even qualitatively judge uncertainty. Thus, this case is not pursued here and the terms uncertainty and risk are treated as synonyms.

A further differentiation can be made with respect to the primary reasons why firms have to face uncertainty. *External uncertainty* results from the complexity and the dynamic of the firm's environment. The conditions on markets for labor, capital, raw material, and products as well as the technological, legal, political, sociological and economic conditions of the firm's environment are subject to permanent changes.[3]

By contrast, *internal uncertainty* accrues from know-how transfers, organizational lacks, or the current and future knowledge and motivation of the firm's staff.

Uncertainty in investment decisions, for example, occurs as price uncertainty for a newly developed product or about the evolution of overall market demand. Uncertainty arises to a high degree in R&D projects for which it is not clear in advance whether the desired result can be obtained or not. Another important differentiation can be made here. On one hand, the firm is exposed to *market (priced) risk*. Market risks denote uncertainties which affect all firms in the same

---

[2]This situation is usually called *risk*.

[3]Increasing complexity can be due to global competition or deregulation. The most important reason for increasing dynamics can be found in the technological breakthroughs of information technology in the last decade.

business. These sources of risk are valued by capital markets. An example is the above mentioned price uncertainty on product and supply markets. On the other hand, there exist so-called *private risks* which contain, for example, the sources of uncertainty in an R&D project. Private risks are solely subject to the individual firm and are not priced by capital markets.[4]

The second important determinant in making investment decisions is *irreversibility*. Irreversibility of investments means the fact that a decision, once being made and implemented, cannot be revised without incurring costs. Although a once built production facility for a certain output good can be closed in response to unfavorable economic conditions, the investment outlay for building the production facility cannot be fully recovered. This is caused by the specificity of economic investments and the inefficiency of secondary markets for investment goods.[5] Investment expenditures are usually firm or industry specific, so that their value for other firms is much less than for the original investor. As a consequence, investment goods, disregarded from their time value, can not be sold for their initial price. For example, most investments in marketing and advertising are firm specific and cannot be recovered. Thus, they can be regarded as completely sunk.

Effect of Managerial Flexibility

In order to counteract existing risks and to limit the associated threat of losses, decision makers have to create and exploit flexibilities embedded in investment projects. Managerial flexibility can therefore be defined as the ability of an economic system to react on unexpected changes inside and outside the system in order to reach a given goal or to reformulate it. In other words, flexibility describes to what degree decision makers are able to adopt new strategies in response to newly arriving information. It is, consequently, a crucial task of management to actively take flexibility into consideration in order to increase the value of the company.

Managerial flexibility can take different forms on different firm levels. It is useful to distinguish *operative*, *strategic* and *financial* flexibility.[6] Operative flexibility refers to opportunities to act contained in a single investment project, e.g., the ability to select a production technology or to temporarily shut down production. Strategic flexibility means the ability to reconfigure the firm´s resource configuration, for example by acquiring other firms, or the opportunity to enter new product markets by successfully completing a previous R&D project. Finally, financial flexibility denotes the ability to manage the capital structure of the firm in order to maximize its value.

Furthermore, flexibility can have a *defensive* or an *offensive* character. An example for the former is the ability to switch to another production input which may protect firms against losses in comparison to being committed to one single input source when prices rise. On the other hand, having the opportunity to expand

---

[4]See Amram/Kulatilaka [7], p. 54.

[5]See Meise [163], p. 10 or similarly Dixit/Pindyck [72], p. 8.

[6]See Koch [115] for a similar differentiation.

to new markets, if demand in the original market exceeds expectations, is obviously offensive in nature.

It turns out that the protective nature of flexibility may limit the downside losses while offensive flexibility can help to fully exploit the upside potential of an investment project in the presence of uncertainty and irreversibility. Either of these two effects introduces an asymmetry in the risk profile of possible investment values, as can be seen in Figure 2.1.

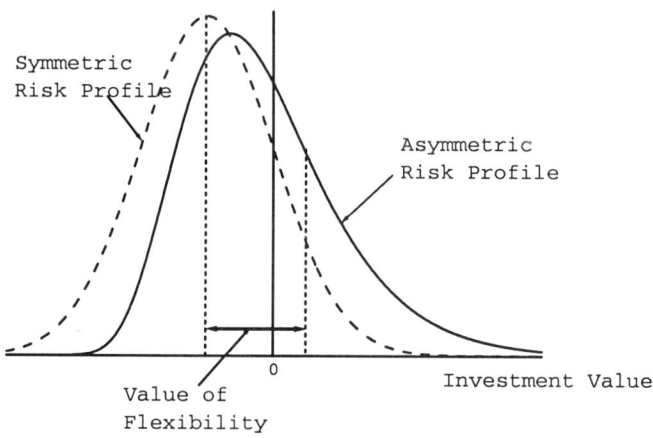

Figure 2.1: Asymmetric risk profile caused by managerial flexibility

The static project, valued as if no flexibility was present, results in a symmetric risk profile with an expected value that is, in this case, less than zero. By considering defensive flexibility, for example the flexibility to abandon the project, the firm is protected against large losses which leads to a left truncation of the investment's risk profile. On the other hand, offensive flexibility allows the firm to take advantage of the existence of uncertainty and irreversibility which results in larger possible project or firm values. Both kinds of flexibility give rise to the risk profile being skewed to the right.

The strategic importance of considering flexibility becomes obvious. In an economic environment characterized by uncertainty and irreversibility, building and exercising flexibility is the crucial task of management in order to enhance the firm's performance and to guarantee its long-term survival. However, in each specific decision situation management has to decide whether the benefit from flexibility exceeds the cost of acquiring it.

In summary, we establish that flexibility is an important part of real investment projects that may not be neglected by management and can add significant value to the firm.

2.1.2   Flexibility and Traditional Capital Budgeting Methods

The basic question to answer in capital budgeting today is whether an investment undertaken by a firm increases overall firm value on capital markets. The underlying idea of market value maximization is due to the *shareholder value* principle which essentially defines the goal of the firm as to exceed the expected rate of return of traded securities with otherwise identical risk structure.[7] Therefore, besides taking into account uncertainty, irreversibility and flexibility, capital budgeting methods should also be consistent with capital market perceptions. In what follows we will shortly discuss why different traditional capital budgeting techniques fail to fulfill all of these four criteria.

Over the past forty years the basic rule for capital budgeting decisions of firms has been the *net present value* (NPV) criterion. Its basic idea is to accept all investment projects for which the sum of the expected, appropriately discounted, future profits, earnings or cash flows are positive and reject the project otherwise. Therefore the *discounted cash flow* (DCF) analysis represents a widely accepted tool for management to calculate the expected NPV and consequently whether a project is favorable or not. Apart from the integration of the *capital asset pricing model* (CAPM)[8] to obtain the risk adjusted discount rate, DCF has stayed almost unchanged in the past two decades. Since the DCF model is useful for evaluating investments that are not related to or dependent on other investment projects of the firm, it produces quite good results if there exist precise estimates of future cash flows. But this implies that the firm faces a clear enough future with little uncertainty and the time horizon of the project is not too long. Since this is usually not the case for most investment decisions, the inaccuracy of future cash flow forecasts can be stated as a serious difficulty in applying DCF.[9]

Furthermore, various defects of DCF analysis have been mentioned by several authors.[10] First, DCF analysis tends to overlook the strategic reasons for an investment. Investing in a project with low profitability to obtain future growth opportunities is not considered by DCF analysis. Consequently, DCF analysis fails to cover the strategic component usually embodied in management's investment decisions, especially in fast growing technology businesses where investments are often undertaken because of competitive reasons rather than short term profit maximization. Second, DCF analysis does not consider the value of flexibility because the NPV criterion boils down to a now or never decision and does not account for management´s ability to adapt for newly arriving information.

---

[7]See Copeland/Koller/Murrin [60], p. 35.

[8]See Sharpe [202], Lintner [145] and Mossin [174].

[9]See Myers [177], p. 133.

[10]See Trigeorgis/Mason [226], Ritchken/Kamrad [191], Kulatilaka/Marcus [128], Ross [193] or Brealey/Myers [31].

At first glance the NPV rule can correctly mimic the idea of capital market valuation. The basic idea of capital market valuation is to use a traded asset with exactly the same cash flow and risk structure as the investment project to be valued in order to determine the market value of the investment. In that, the NPV criterion is consistent with modern capital market theory. However, it values investments as if firms were committed to take its actions along a certain expected scenario. In essence, the NPV approach assumes that management makes at the beginning an irrevocable commitment to an operating strategy from which it cannot depart; regardless whether the firm´s uncertain environment remains faithful to or deviates from the expected scenario of cash flows. Consequently, the firm is assumed to be exposed to uncertainty symmetrically because it neither can profit from favorable developments nor limit its losses if circumstances turn out bad. It follows that the NPV criterion cannot include the asymmetry introduced by managerial flexibility. It undervalues investment projects that contain flexibility in the presence of uncertainty and irreversibility and can lead to wrong investment decisions.

Decision tree analysis (DTA) is an extension of the NPV approach to overcome the lack of taking account for flexibility. Again uncertainty over future cash flows is taken care of by using risk adjusted discount rates to determine today´s expected value. With DTA, flexibility is considered by decomposing the investment project into a sequence of decisions and states of the firm's uncertain environment. The resulting tree-like structure of states and decisions can be used to determine the (risk adjusted) value of an investment project and the corresponding optimal decision strategy the firm should adopt in response to changes in its uncertain environment. As such it is particularly useful for analyzing complex sequential investment decisions when uncertainty is resolved at distinct points in time. Whereas conventional NPV analysis might be misused by managers inclined to focus only on the initial decision to accept or reject the project at the expense of subsequent decisions dependent on it, DTA forces management to recognize explicitly the interdependencies between the initial decision to invest and subsequent decisions.

DTA is, in principle, able to accommodate managerial flexibility into the capital budgeting process. However, there exists a serious problem in how to determine the appropriate discount rate. Using a constant risk adjusted discount rate (derived for example from the CAPM) presumes that the risk borne per period is constant and that uncertainty is resolved continuously at a constant rate over time. But this assumption is not met by DTA. In fact, at each point in time where either an uncertain event happens or a managerial action is taken, the distribution over future cash flows is altered and with that the risk structure of the project as well. Therefore, in order to correctly value the project in DTA we would have to find a different risk adjusted discount rate for each time period, for each possible state of nature and for each possible managerial strategy. While the stochastic environment of the firm, in principle, evolves continuously, the problem of finding an appropriate discount rate is even aggravated because the risk adjustment of the discount rate has to be continuous as well. We therefore state that DTA cannot be used to correctly value investment projects in the presence of uncertainty, irreversibility and managerial flexibility.[11]

In summary, it can be concluded that the traditional valuation methods, among which we briefly discussed the net present value analysis and the decision tree analysis, are not capable of dealing with managerial flexibility in a theoretically correct manner. Following these methods when making investment decisions may lead to tremendous mistakes.[12]

2.1.3 Towards a New Investment Paradigm

In order to overcome the obstacles of traditional valuation methods the *real option valuation approach* has emerged during the last two decades.[13] The basic idea of valuing managerial flexibility in the real option approach is to interpret flexibility as options on real assets which can be valued similarly to financial options traded at capital markets.

The real option idea rests on three pillars. First, in viewing managerial flexibility as options, the objective function in comparison to the (static) net present value approach has to be extended by an option premium representing the (market) value of flexibility in addition to the static project value. Second, the analogy between real and financial options and its limitations has to be established. And third, the assumptions that have to be satisfied in order to value real options as financial options need to be clarified. We will shortly work through each of these points.

Expanded Net Present Value

As already discussed the NPV approach is only capable of valuing investment projects with either little uncertainty or no managerial flexibility. Therefore, the net present value of a project can be seen as the value of the basis investment which needs to be extended by the value of flexibility to limit downside losses and to exploit upside profits. Since flexibility provides management with the right, but not the obligation, to exercise the project inherent flexibility — the option like features become obvious, an option premium needs to be added to the static net present value. The true value of an investment project therefore consists of an *expanded net present value* (ENPV) which is defined as:

$$\text{ENPV} \quad = \quad \text{Static NPV} + \text{Value of Real Options (Option Premium)}.$$

---

[11]The same arguments apply to any kind of stochastic dynamic programming approach, see Dixit/Pindyck [72], p. 120, or Smith/Nau [211], p. 799, and Smith/McCardle [209], p. 201, for decision analysis.

[12]McDonald [161] examined the valuation error caused by using the wrong capital budgeting method to value managerial flexibility and found that the differences can be substantial.

[13]The term *real options* goes back to Myers [176]. Prerequisite for the development of real option pricing theory were the advances in the theory of financial options introduced by the nobel prize winning authors Myron Scholes and Robert C. Merton and the late Fisher Black in their seminal works Black/Scholes [27] and Merton [165].

Consequently, the NPV rule keeps its principle justification to assess investment projects. However, it needs to be extended by a dynamic component containing the embedded real options. Because option values are always nonnegative, the static NPV rule can lead to an undervaluation of investment projects of firms. It can even be useful to undertake an investment project although its static NPV is negative if the option premium is large enough to yield a positive ENPV.

Similarity to Financial Options

In order to solve the main question of how flexibility can be valued quantitatively, we can fall back on the analogy between financial and real investments which is the foundation to determine the static NPV as well. If financial contracts (or contingent claims) exist which have the same risk structure as the real options associated with managerial flexibility, then the models developed to value these contingent claims can also be applied to the valuation of flexibility. From that point of view, it is obvious to interpret flexibility as options on real assets and try to value them using financial option pricing theory (more generally contingent claims analysis).

Although we do not go into the mathematical details of option pricing theory,[14] the main properties of real options and their close resemblance to financial options will be discussed.

A financial option can be generally defined as a contract which gives its holder the right, but not the obligation, to exchange a financial contract for another one under predetermined conditions. For example, a call option written on a certain stock provides its holder with the right to *purchase* the stock, referred to as *underlying asset*, in exchange for a certain predetermined price, referred to as *exercise price*. Vice versa, a put option gives its holder the right to *sell* the underlying asset for a certain price. If the right to exchange the two assets is fixed to a certain *exercise date*, the option is of *European* type, while it is called of *American* type if the option can be exercised throughout the time interval to expiration.

The value of any financial option contains two parts. The first part is the so-called *intrinsic value* which denotes the value of the option if it were immediately exercised. That is, for a call option the current market value of the underlying asset net of the exercise price if the underlying asset is larger than the exercise price and zero otherwise, since no rational investor would use his right to exchange the two assets in that case. The second component contains the *time value* of the option. It represents the difference between the current value of the option and its intrinsic value. The time value of an option is always positive. The reason is that the time value of an option reflects the possibility that the stochastic movements of the underlying asset result in an even higher intrinsic value at the expiration date of the option. Of course, at the expiration date the time value of the option is zero, and the value of the option coincides with its intrinsic value.

---

[14]The mathematical foundations of option pricing theory will be presented in subsequent chapters. For an introduction to financial option pricing theory see, e.g., Brealey/Myers [31], Hull [102], Duffie [75], Karatzas/Shreve [109] in ascending order of mathematical rigor.

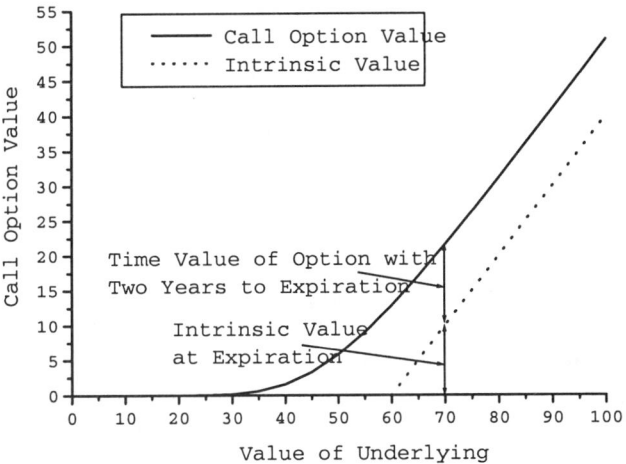

Figure 2.2: Time value and intrinsic value of a European call option with two years to expiration.

Figure 2.2 shows the typical asymmetry in the payoff structure of an option. The example shows the value of a European call option as a function of a non-dividend paying underlying asset S with exercise price $X = 60$ and a time to expiration of $T = 2$ years. For the calculation it was assumed that the instantaneous constant interest rate r is 10% p.a. and that the uncertainty in the price movements of the underlying asset can be measured through a volatility σ of 20% p.a.. The intrinsic value of a European call option is given by the payoff it delivers at the expiration date. Because the option need not be exercised if the value of the underlying asset S is less than the exercise price X, the terminal payoff of the option is

$$C_T = \max\{S - X, 0\}.$$

It can be seen from the figure that the value of the option can be significantly higher than its intrinsic value and that it is always nonnegative, since the loss for the holder is limited due to the ability to let the option expire unexercised.

The conceptual analogy between financial options and managerial flexibility rests on the similarity of their payoff and risk structure. As with a financial option the firm has the right, but not the obligation, to exercise its flexibility and may profit from good developments while being protected against bad.

A firm facing an investment decision therefore possesses the *option to invest*.[15] It need not exercise this option immediately but can rather wait until circumstances turn out to be good enough to undertake the investment — or let the option expire unexercised otherwise. In option pricing terms the firm has the option to acquire the investment project with a net present value (underlying asset) up to a certain expiration date for paying the investment cost (exercise price) necessary to install the project. Taking this real option into account results in just the same asymmetric payoff structure as observed for the financial option in Figure 2.2.

The figure also displays the major failure of the (static) NPV approach. Using the NPV rule the option to invest is valued as if the investor *has to exercise* the option immediately. If the option is in-the-money, i.e., its static NPV is positive, applying the (static) NPV rule recovers the positive intrinsic value of the option but kills its time value, leading to the irrational decision of immediately investing in the project. If on the other hand the option is out-of-the-money, i.e., its static NPV is negative, the (static) NPV rule proclaims to once and forever reject the investment project under consideration, neglecting the fact that the investment opportunity itself may still have a positive value.

	Call option on a stock	Option to invest
Underlying asset	Current stock value	Static NPV of cash flows
Exercise price	Fixed stock price	Investment cost
Time of expiration	Exercise date	Time until opportunity disappears
Risk	Stock value uncertainty	Project value uncertainty
Interest rate	Riskless interest rate	Riskless interest rate
Dividend payments	Payments to stock holders	Payments lost through waiting

Table 2.1: Comparison of the option analogy between a call option on a stock and the option to invest

Table 2.1[16] summarizes the analogy between a financial call option on a stock and the option to invest in a (real) investment project. However, the close analogy between financial and real options is not quite complete. Although real options can, like financial options, be specific contracts, they are usually rather opportunities to take a specific action. This implies that the parameters which determine the

---

[15]See McDonald/Siegel [160].
[16]See Trigeorgis [225], p. 125.

value of the option (expiration date, exercise price etc.) are not necessarily fixed. Finally, the term *option on a real asset* has to be specified. The value of an asset is determined by the present value of the future cash flows generated by this asset. This implies that a sequence of future cash flows is connected to each asset.

After these preliminary reflections the term *real option* can be defined as follows: Real options represent opportunities to act which provide their holder with the right, within the scope of a certain transaction, to exchange the value of the cash flow stream of an underlying asset against the value of the cash flow stream of an exercise asset.

This definition shows that in principle a firm consists of a portfolio of real options which management should value and optimally exercise.[17]

In the literature several kinds of flexibilities have been identified where the options analogy can be applied.[18] Beyond the option to invest these are the:

- option to defer[19]

- option to abandon[20]

- option to alter operating scale[21]

- option to switch (inputs or outputs)[22]

- time-to-build option[23]

- growth option[24]

If the firm has the ability to defer (or wait) undertaking the investment project, it holds a call option on the project which need not be exercised immediately. The crucial task is to determine when the project value is high enough to justify making the irreversible investment.[25] In deferring the implementation of the project the firm has to weigh the benefit of waiting (learning about the future development) against the opportunity cost of waiting (foregone cash flows while not being invested).

---

[17]See Luehrman [148].

[18]See, e.g., Trigeorgis [225], p. 2.

[19]See Tourinho [218], Titman [217], McDonald/Siegel [160], Paddock/Siegel/Smith [182], Pindyck [186], Majd/Pindyck [151], Ingersoll/Ross [104], Metcalf/Hassett [167], Alvarez [4], Dixit/Pindyck/Sødal [74], Höger [97], Sødal [213].

[20]See Myers/Majd [178], Dixit [71], Alvarez [5].

[21]See Brennan/Schwartz [34], McDonald/Siegel [159], Trigeorgis/Mason [226], Pindyck [186], Dixit [69], He/Pindyck [93], Dixit/Pindyck [72].

[22]See Margrabe [153], Kulatilaka [123], Kulatilaka [124], Kulatilaka/Trigeorgis [132], Kamrad/Ernst [107], Carr [44], Vollert [229].

[23]See Majd/Pindyck [150], Carr [43], Pindyck [187], Bar-Ilan/Strange [12, 13, 14].

[24]Myers [176], Kester [110, 111], Chung/Chaorenwong [52], Kogut [116], Kogut/Kulatilaka [118], Smith/Triantis [212], Willner [233], Chung/Kim [53], Kulatilaka/Perotti [130], Ottoo [181], Berk/Green/Naik [24], Beliossi/Smit [18], Koch [115].

[25]See Dixit/Pindyck [72], chapter 4, for a good introduction to the optimal exercise of the option to defer.

Similarly, the option to abandon a project and sell it for its salvage value can be exercised if market conditions decline severely. By holding the risky project and having the option to recover a certain part of the initial investment sum, the option to abandon possesses the properties of a put option on a dividend paying stock.

In contrast to the above mentioned real options, the option to alter the operating scale can have different directions. It covers cases where the firm has the opportunity to expand or contract its scale or even completely shut down and reopen its facilities. For example, the option to expand can be seen as a call option on additional capacity, while the option to contract admits put option-like features.

So far, the exercise price of the real options was assumed to be known with certainty. However, if the option to switch, for example production inputs or outputs, is considered, the firm has the option to exchange one risky asset for another, which means that the exercise price of the option is now stochastic. Then, it is not clear whether the option to switch should be interpreted as a call or put option.[26]

Another source of flexibility can be due to staging investment. Especially, in capital intensive industries it can be useful not to invest the whole initial outlay at once, but to complete the investment project step by step and abandon it in midstream if new information is unfavorable. The time-to-build option can be used to evaluate the advantage of staging investment. Since the completion of each new step gives the firm the right to start with the next step, investing in stages can be interpreted as call options on call options. Such option compoundness can be valued using the compound option approach of Geske [81].[27]

While all the above mentioned real options refer to managerial flexibility contained in one and the same project, growth options deal with interproject compoundness. "An early investment (e.g., R&D, lease on undeveloped land or oil reserves, strategic acquisition) is a prerequisite or a link in a chain of interrelated projects opening up future growth opportunities (e.g., new product or process, oil reserves, access to new markets, strengthening of core capabilities)."[28] These early strategic investments possess again the nature of compound options, since the option to invest in the early project is a call option on the option to invest in the subsequent project. Therefore growth options focus on the strategic nature of investments.[29] [30]

---

[26]See Margrabe [153].

[27]It is worth noting that a financial call option is, in principle, nothing else than a call option on the stock of a firm which in itself can be interpreted as a call option on the firm value due to its limited liability. Therefore a financial option can also be viewed as a compound option. This point was also made by Geske [80].

[28]Trigeorgis [225], p. 3.

[29]In fact, the seminal paper of Myers [176] exclusively dealt with growth options as a source of additional firm value.

[30]Growth options are especially useful for assessing the value of R&D projects, see Huchzermeier/Loch [101], venture capital investments, see Willner [233], joint ventures, see Kogut [116] and Lee [142], and similar strategic investments. Furthermore, the growth options idea can be used to explain the phenomenon that high growth companies, e.g., internet firms, are assigned an extremely large market value although they will not earn any profits in the near future, see Mauboussin [155].

It can be concluded from the above examples that almost any kind of managerial flexibility can be interpreted as a real option, and that the analogy between financial and real options holds at least on a conceptualized level. In order to be actually able to quantify managerial flexibility as real options, it has to be checked whether the assumptions made for financial option pricing are also met by real options.

Assumptions for Valuing Real Options as Financial Options

The ability to value managerial flexibility as real options depends on the question of what assumptions need to be satisfied in order to apply financial option pricing theory.

The assumptions for valuing financial options are the following:[31]

- Markets are complete.

- Markets are frictionless (for stocks, bonds, and options), i.e., there are no transaction costs and no restrictions on short sales; all securities are infinitely divisible; and borrowing and lending (at the same rate) is unrestricted.

- The risk free (short-term) interest rate is constant over the life of the option (or known over time).

- Any dividends are known in size and date of payment.

- The underlying asset follows a known stochastic process.

- Investors are rational and prefer more to less.

If these assumptions are met, the basic idea of financial option pricing is that the seller of the option can mimic the payoff structure of the option in any state of the world by investing the amount obtained by selling the option in a *replicating portfolio* consisting of (riskless) bonds and the underlying asset. Therefore the overall portfolio, consisting of the current market price of the option and the replicating portfolio, has always zero risk. In the absence of *arbitrage opportunities*[32] the replicating portfolio consequently earns the riskless interest rate. The problem of valuing the option is therefore transformed into the *risk-neutral world*[33] where discounting using the risk free interest rate is possible. The beauty of the approach is that in the risk-neutral world all investors are preference free, which means that any investor regardless of his risk preferences assigns the same value to the option and therefore determines the option's unique market price.

---

[31] See Black/Scholes [27], p. 640.

[32] See Brealey/Myers [31], p. 36, for an intuitive explanation, Hull [102], p. 236, for the meaning of the no arbitrage assumption for pricing options and Karatzas/Shreve [109], p. 327ff, for a mathematical presentation.

[33] See Harrison/Kreps [88].

In order to set up the replicating portfolio it is necessary to continuously re-balance it because, with each change in the stochastically evolving price of the underlying asset, the value of the option changes as well. This is assured by the second assumption which makes continuous trading and therefore continuous re-balancing feasible.

The most crucial assumption for the valuation of financial options concerns market completeness. Markets are said to be complete if there exist at least as many uncorrelated primary securities (portfolios of stocks and bonds) as sources of risk. If this assumption is violated, there may exist contingent claims (e.g., options) whose risk cannot be completely hedged by constructing a replicating portfolio. Then, the overall portfolio of option and replicating portfolio bears some remaining risk towards which the perception of investors on capital markets cannot be deter-mined without considering the individual investors' risk preferences. Therefore, in incomplete markets the preference free, market oriented approach which leads to one single rational option price fails to apply. In incomplete markets option prices cannot be interpreted as market values but rather as lower limits.[34]

Of course, the same assumptions that need to be satisfied for financial options should hold for valuing real options in order to determine the market value of man-agerial flexibility. The assumption of complete markets is again of utter importance. For real option pricing, market completeness means that the *spanning condition* holds. The latter requires that the underlying asset of the real option or another perfectly correlated portfolio of securities need to be traded at capital markets such that the risk structure of the real option's underlying can be *duplicated*.[35] The ful-fillment of the spanning condition ensures that the market value of the replicated portfolio, and therefore the value of the real option, can be determined.

Although real options and the real options' underlying assets are often only infrequently or not at all traded at capital markets, this does not mean that the spanning condition is violated. It is sufficient to find a traded asset that possesses the same risk structure as the investment project containing the real options. Real option valuation is therefore not restricted by the lack of liquid markets for real options. The main problem with real options is rather to find and identify traded assets or portfolios of traded assets with the same risk structure as the real option's underlying. Furthermore, the real option's underlying source of risk might not even exist in the present as long as markets are complete, as is for example the case for the option to defer where the stochastic underlying project value does not yet exist.[36]

This implies that the application of real option pricing is not limited by capital market assumptions but rather by problems in identifying a traded asset — so-called *twin asset* — with the same risk structure as the real option's underlying asset.

---

[34]A review of the theory of *incomplete financial markets* can for example be found in Bingham/Kiesel [26] or Karatzas/Shreve [109].

[35]See Meise [163], p. 60.

[36]See Sick [205], p. 652.

Yet, even if the spanning condition is not satisfied and real option pricing theory fails to apply, so do all other market value oriented capital budgeting methods, among others the NPV approach and decision tree analysis. The reason is that any modern capital budgeting method tries to circumvent the problem of preference dependent valuation by building the bridge to capital markets in order to derive market values of investment projects. These methods also rest on the assumption of complete capital markets just like the real option approach.

For example the NPV approach needs the spanning condition, and implicitly market completeness, in order to be able to fall back on an asset with identical risk structure, from which it determines the relevant risk adjusted discount rate. The real option approach in turn values an option inherent in an investment project by comparing it with the market value a financial option written on the traded twin asset would have. As a consequence, both approaches rest on exactly the same assumptions.

To conclude, there exist no principle objections to using the capital market analogy to evaluate real options as long as the main differences between the two approaches are taken into account. Since real option pricing rests on the same assumptions as traditional capital budgeting methods, like DCF, it represents an important extension of the latter that is able to correctly value managerial flexibility in the presence of uncertainty and irreversibility.

2.2 Classification of Real Options

In order to value and manage the real options associated with flexibility it is important to classify the real options that occur in managerial practice and to discuss the special features that arise when valuing real options. We therefore develop classification frameworks for real options from a management perspective and a valuation perspective. Although both dimensions are closely connected, this differentiation allows us to stress the economic implications of the real option approach without neglecting the specifics of the valuation process. It is especially important to keep both perspectives in mind in order to build a bridge between the management perspective of flexibility, which is strongly application oriented, and the valuation tools for real options, which are due to financial option pricing theory.

2.2.1 Management Perspective

In 1977, Myers [176] suggested that the market value of a firm consists of the present value of its expected cash flows *plus* the value of the firm's opportunities for growth,[37] which Myers described as the firm's *real options*. Following this insight, Baldwin/Mason/Ruback [11] recognized that a firm's growth opportunities may have various degrees of flexibility associated with them. They termed as *operating*

---

[37]Note that this is only part of our definition of the expanded net present value following Trigeorgis [220], p. 123.

options the choices that determine the operational flexibility associated with each of the firm's opportunities for growth. In the sequel, the term *real options* has become generic for both the basic opportunities for growth defined by Myers as well as the operating options noted by Baldwin et al.

However, this definition is not quite complete and we will follow a different scheme in this book. Managerial flexibility can be related to either future investment projects — *strategic real options* — or already undertaken investment projects, i.e., already existing assets — *operative real options*. While both kinds of real options refer to the assets of the firm, *financial real options* refer to the flexibility of the firm's capital.[38]

Using these definitions it is possible to further decompose the expanded net present value of a firm into four basic components, as displayed in Figure 2.3.

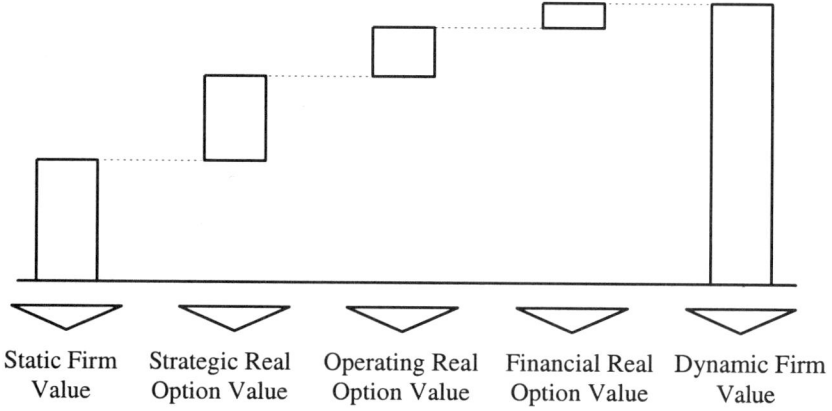

| Static Firm | Strategic Real | Operating Real | Financial Real | Dynamic Firm |
| Value | Option Value | Option Value | Option Value | Value |

Figure 2.3: Sources of firm value

The different sources of value that determine the *dynamic value* of the firm will be discussed shortly.

While the static firm value is determined by its *assets in place*, it can be valued using standard discounted cash flow analysis with risk adjusted discount rates. The estimation of the future free cash flows is performed as if the firm adopts its originally chosen strategy without taking flexibility into account.

Strategic real options reflect the value of managerial flexibility to undertake additional investments or disinvestments in the future in order to create new *strategic assets*. Strategic assets can be tangible assets like production facilities, but also intangible assets like organizational capabilities, the firm's knowledge base, market position etc.. Therefore strategic options can be understood as the flexibility held or acquired by the firm to create and exploit its future business opportunities.[39]

---

[38]Financial real options are not to be confused with viewing the equity of the firm as an option on the value of the firm as in Geske [81].

Examples for strategic options are:

- growth options

- M&A options[40]

- joint venture options[41]

- strategic alliances and insurances[42]

- disinvestment options[43]

Strategic real options can amount to a significant part of overall firm value.[44]

In recent years the close resemblance between the real options approach and the resource based view of the firm in strategic management has been recognized. In a wider sense strategic resources can also be interpreted as growth options which admit transformation of the firm's capabilities into sustainable competitive advantage.[45] Although we will not pursue this idea any further, its pure existence highlights the strategic importance of the real options approach beyond its application as a valuation tool.

In contrast to strategic options, operating real options refer to the value of flexibility due to active management of existing assets. That means, the firm's operating strategy can be adjusted in response to the gradually resolving uncertainty and with it a positive influence on the future cash flows can be achieved. Operating real options do not create new assets but improve the returns from existing assets.[46] Furthermore, operating real options may be naturally *embedded* into a project, e.g., the option to temporarily shut down, or must be *acquired* while carrying out the investment, e.g., the option to switch outputs in a flexible manufacturing system.

Operating real options can be further distinguished in:[47]

---

[39]For a treatment of the importance of the real option approach for strategic management see, e.g., Sharp [201], Nichols [179], Dixit/Pindyck [73], Courtney/Kirkland/Viguerie [65], Leslie/Michaels [143], Copeland/Keenan [59, 58], Luehrman [147, 148], Amram/Kulatilaka [6, 8], Beinhocker [17].

[40]See Lee [142], p. 118; and similar Koch [115], p. 95.

[41]See Lee [142], p. 79.

[42]See Lee [142], p. 165.

[43]See Myers/Majd [178], Berger/Ofek/Swary [22], p. 259.

[44]See Kester [110], Beliossi/Smit [18] for empirical results.

[45]Although it is hard to value the strategic resources of a firm using financial option pricing theory, the idea of merging real options and strategic management theory is very promising on a conceptualized level. The early precursors of this newly emerging branch of research are Kogut/Kulatilaka [118], Smith/Triantis [212], Sanchez [197], Lau [140], Foss [79], Kogut/Kulatilaka [119], Kulatilaka/Venkatraman [133], Bernardo/Chowdry [25], Williamson [232].

[46]The distinction between strategic and operative real options is not clearly cut. For example, the flexibility to expand production can be either seen as an operative option because it alters the existing production facility or it may be seen as a strategic real option because expansion of a production facility creates a new asset.

[47]See Koch [115], p. 166.

- input options

- process options

- output options

- expansion/contraction options

Input options deal with the flexibility to switch, for example, between different suppliers or different input goods.[48] Process options allow management to choose among different production[49] or supply technologies[50] while output options determine the value of having the flexibility to switch between different products in manufacturing and commercialization.[51] Finally, options to expand and contract can be applied to react flexibly on changes in demand by purchasing or selling additional capacity.[52]

The creation and management of operating real options is aimed at an efficient utilization of the firm's resources. Like strategic real options, operating real options admit the typical skewness due to option asymmetry. There exists often a close connection between operating and strategic real options because being highly flexible with respect to existing real assets in many cases implies possessing flexibility with respect to future strategic assets.

The final source of option value is in the ability to choose and change the capital structure of the firm during the course of a project. There are basically two sources of option value in financing decisions. The first is due to the option-like features of equity because of its limited liability.[53] The second comes from the flexibility of recapitalization,[54] i.e., issuing new stocks, repurchasing outstanding stocks, debt financing of new ventures or changing the debt maturity etc. Increasing the debt ratio can yield, for example, tax shield advantages when the firm is currently earning high profits due to favorable market conditions. Considering financial real options can have several important implications. Despite the finding of Mauer/Triantis [157] that the ability to recapitalize during a certain project also containing other operating real options does, in essence, not change the optimal operating policy, they found that financial flexibility can add additional value to

---

[48] See Kulatilaka [124].

[49] See Grenadier/Weiss [85].

[50] See Cohen/Huchzermeier [54, 55].

[51] See Triantis/Hodder [219], Hodder/Triantis [96] and Sanchez [196].

[52] See McLaughlin/Taggert [162].

[53] See Trigeorgis [224], p. 216.

[54] See Mauer/Triantis [157], Lambrecht [136].

the firm.[55] Furthermore, there may arise agency problems with debt financing[56] The optimal strategy adopted by the firm is different if it exercises its real option such that *equity value* is maximized rather than *overall firm value including debt.*

Although the classification from a management perspective covers the main managerial flexibilities that can be modelled as real options, the decomposition of firm value into the four components should not be understood as a sum of four separate values. The overall value of different real options contained in one and the same project or firm are typically not additive. Therefore the decomposition of firm value should be rather seen as a concept to allocate the main sources of firm value and to assign the responsibilities for the management of different real options within an organizational structure.

2.2.2 Valuation Perspective

The valuation perspective of real options is concerned with the specific features of real options that need to be considered when applying financial option pricing theory. Two different topics can be distinguished when discussing the valuation perspective. The first direction aims at the differences concerning the input parameters for real and financial option pricing theory. The second deals with the limitations of the option analogy. Due to economic reasoning there exist certain differences (e.g., the presence of competition) between real options and financial options. These differences can be exploited to develop a classification scheme for real options that characterizes its main properties from a valuation perspective.

Characteristics of the Input Parameters

Starting at the financial option analogy it was demonstrated in Section 2.1.3 that most real and financial options can be characterized by six input variables.[57]

The first important inputs for a real option analysis are the several sources of uncertainty that characterize the firm's uncertain environment. Figure 2.4 demonstrates several sources of risk to which the firm may be exposed. While most sources of uncertainty arise from the firm's economic environment there are other sources of risk that may seriously affect the firm's profitability as well. Technological risk can, for example, be especially important in R&D decisions.[58] Timing risk may occur in situations where large investments need to be completed at a

---

[55]Mauer/Triantis wrote: "The lack of an economically significant effect of debt financing on the investment decision has an important practical implication for capital budgeting decisions. Specifically, it suggests that a levered firm may make its investment timing decisions ignoring the influence of debt financing, thus greatly simplifying the investment decision," see Mauer/Triantis [157], p. 1270. By way of contrast, Lambrecht [136], p. 27, argues that in a duopolistic setting the capital structure and the ability to adjust the capital structure can have significant impact on firms' investment decisions. It seems that this problem is context and modelling specific and cannot be answered in general.

[56]See Mauer/Ott [156], p. 153.

[57]See Table 2.1.

[58]See e.g. Pennings/Lint [185], Weeds [230].

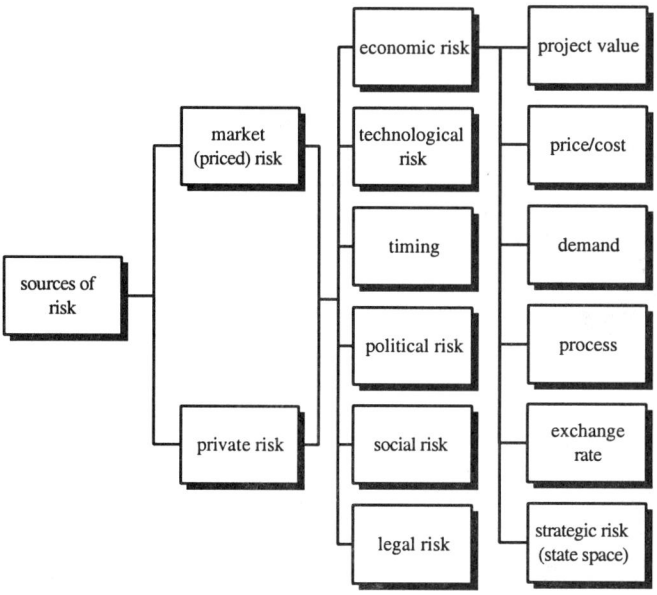

Figure 2.4: Different sources of risk

certain date.[59] And political, legal and social risks can be crucial to, e.g., foreign investment decisions[60] or industries which are exposed to severe governmental regulations.[61] All of the just mentioned sources of risk are usually combined with a certain economic risk that measures the profitability of the project directly or indirectly. Modelling the project value itself as underlying source of uncertainty for the option valuation is certainly the most often used alternative.[62] While there exist some principle disadvantages of using the project value itself as underlying source of risk,[63] a decomposition into price, cost, demand, process, exchange rate, or strategic risk — to name a few — is more appropriate. Considering price or cost risk is especially suitable for decision situations where the input and output goods are traded at liquid markets, as is the case for example for decisions in the natural resources industry.[64] On the other hand, in industries where the price is

---

[59] See e.g., Liu/Yao [146].

[60] See e.g., Fischer [78].

[61] See e.g., Mason/Baldwin [154], Teisberg [216].

[62] See Dixit/Pindyck [72] and Trigeorgis [225].

[63] See the discussion in footnote 27 of Chapter 3.

[64] See e.g., Brennan/Schwartz [34], Smit [207], Dias/Rocha [68], Lund [149] and the extensive

itself a strategic variable it can be useful to consider overall market demand uncertainty and the elasticity of demand with respect to price changes.[65] Process risk can be modelled when there exists output yield uncertainty[66] or uncertainty about the quality level of the output good.[67] Exchange rate risk is important for foreign investment decisions.[68] Finally, it might be worthwhile to consider strategic risk which is caused by two effects. First, several sources of uncertainty that are important for decision making are only infrequently observable or only with serious errors.[69] Second, there might exist uncertainty about the future real options that are embedded in a new project, which gives rise to uncertainty over the future state space or strategy space the firm may navigate in.

Of course, in real life decision making many of these risks are present at the same time and it is a difficult task to model all of them at once. In practical applications decision makers should concentrate on a few sources of uncertainty which most seriously affect the success of an investment project. However, even with a limited number of considered uncertainties it is still a very difficult empirical question to establish the capital market link by finding a spanning asset that mimics the risk structure of the source of uncertainty to be modeled.[70] Furthermore, many sources of uncertainty, like technological risk, are completely due to the firm and are not priced by the capital market. This poses two problems. First, it is difficult to determine which sources of risk that the firm is exposed to are market priced and which are not.[71] Second, if the source of risk is not market priced, i.e., if it is private to the firm (see Figure 2.4), then financial option pricing fails to apply towards this kind of risk as underlying, and we would, in principle, need to know the risk preferences of management towards this risk.[72] [73]

Provided these principle obstacles are overcome, the next important question is which mathematical framework to choose in order to model the underlying sources of risk. Basically, capital market valuation models can be either time discrete or continuous. Furthermore, the kind of stochastic process chosen to represent the time evolution of the underlying source of risk is crucial to the results of the real

---

number of references therein.

[65] See e.g., Pindyck [186], Alvarez [4], Metters [168], Cohen/Huchzermeier [55].

[66] see e.g., Kamrad/Ernst [107].

[67] See e.g., Sagi [195].

[68] See e.g., Capel [41], Kogut/Kulatilaka [117], Bell [19], Mello/Parsons/Triantis [164], Huchzermeier/Cohen [100] and the book of Buckley [36].

[69] See e.g., Childs/Ott/Riddiough [49]

[70] In fact, to date there exists to our knowledge no empirical work that relates sources of risk more abstract than prices for natural resources, exchange rates or other variables for which directly or indirectly traded financial contracts exist to traded assets on capital markets.

[71] See Amram/Kulatilaka [7], p. 56

[72] In the sequel of this book it is assumed for simplicity that all investors are risk-neutral towards private risk.

[73] See Moel/Tufano [172] p. 7, for an example of the treatment of private and market priced risk at the same time in the mining industry. Similarly, see Berk/Green/Naik [23] for private versus market risk in R&D ventures. For the general treatment of private risk in a decision analysis framework see Smith/McCardle [210, 209] and Smith/Nau [211].

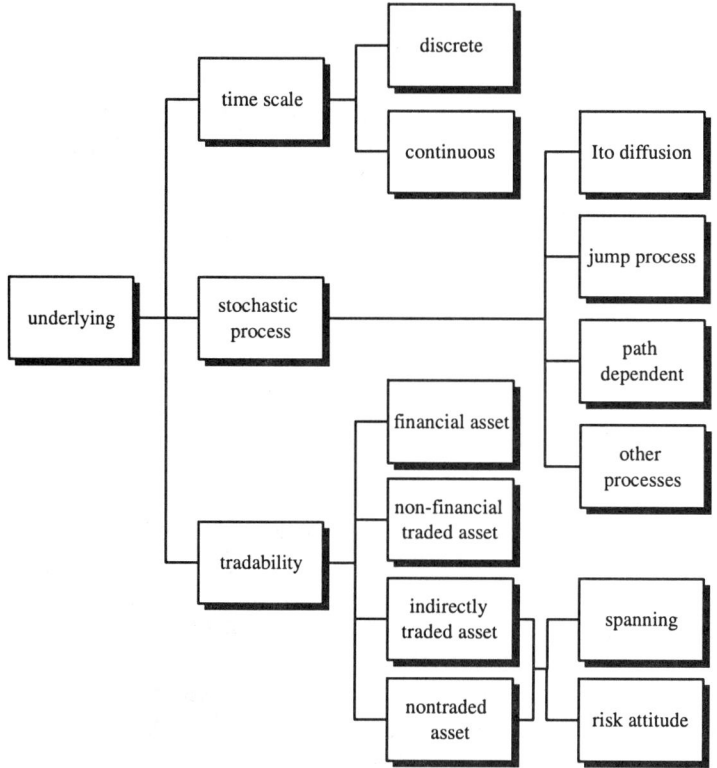

Figure 2.5: Characteristics of the underlying

option valuation. In classical mathematical finance, market prices of stocks are usually modelled as continuous diffusions where the risk comes into the model through Brownian motions well known from physics. The general class of models covering these processes is called Itô diffusion.[74] Sometimes a continuous path process is not sufficient for modelling asset price processes because suddenly arriving information may force the price process to exhibit jumps. Discontinuities in asset price processes can be represented by jump processes.[75] Jump processes can for example be useful to model the uncertainty of a technological breakthrough[76] or the random arrival of competitors.[77] While the two above mentioned stochastic processes are in its original form memoryless,[78] due to for example seasonal

---

[74] See Dixit/Pindyck [72], p. 70.

[75] The mathematical exposition of Itô diffusions and jump processes is given in the following chapter.

[76] See e.g., Pennings [184] for an example of jump processes in R&D, Willner [233] for venture capital, or Farzin/Huismann/Kort [77] for technology adoption.

[77] See Dixit/Pindyck [72], p. 167, or Trigeorgis [225], p. 284.

effects, the sources of uncertainty may also follow path-dependent processes, i.e., stochastic processes that depend on the history of the process.[79] Although these three different stochastic processes are very flexible in modelling nearly all possible price path evolutions that are relevant for real and financial option pricing, recent empirical findings suggest that asset price processes are sometimes more realistically modelled using so-called stable distributions and Levy-motions instead of the Brownian motions and jump processes.[80]

The final topic concerning the underlying is whether the underlying source of uncertainty is traded or not. If it is a financial or non-financial traded asset, there exists no principle problem in applying real option pricing besides identifying the correct stochastic process governing its evolution over time. However, for non-traded underlying sources of uncertainty, real option values can only be determined if the spanning condition holds. Otherwise there is no way but to employ a decision analysis oriented valuation procedure which considers individual investors' risk preferences and utility functions in order to derive subjective values for the firm's real options.[81]

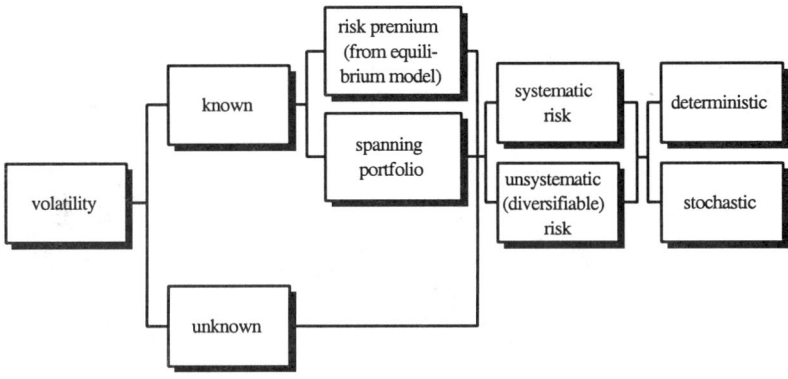

Figure 2.6: Characteristics of the volatility

The risk inherent in the stochastic process of the underlying source of uncertainty is often measured through the volatility σ. Although it might be difficult to estimate the volatility, there is often market or historical information available that

---

[78] In mathematical terms they admit the strong Markov property.

[79] See e.g., Cherian/Patel/Khripko [48].

[80] These empirical findings go back to Mandelbrot/Taylor [152] who observed heavy-tailedness in return distributions of financial assets that cannot be recovered using Brownian motions. In recent years various authors have taken up and extended this idea, see e.g., Mittnik/Rachev [170], Mittnik/Rachev/Paolella [171]. In real option pricing only two works have been published dealing with the subject of stable distributions, see Vollert [229] and Boyarchenko/Levendorskiĭ [30].

[81] See Lee [142] and Kulatilaka [126, 125].

helps at least to roughly determine the relevant range of values for the volatility.[82]
If a spanning portfolio can be found, then the problem reduces to estimating the
volatility of the spanning portfolio, for which there is plenty of data available. If,
on the other hand, a spanning portfolio cannot be identified for each single source
of risk, it is necessary to choose a risk premium that reflects the market price of
the whole bundle of the firm's risks which can, for example, be obtained by an
equilibrium approach such as the CAPM.

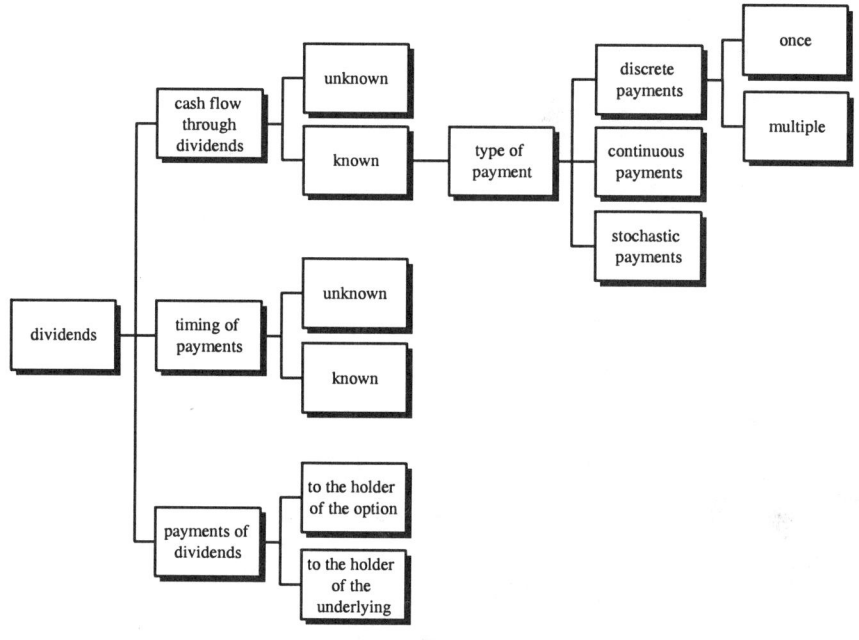

Figure 2.7: Characteristics of dividends

Any cash flow, like payment that arises in the context of an investment project
can be modelled as a dividend payment in financial option pricing theory. When it
comes to dividend payments the amount and frequency is important. The amount of
the dividend can be either deterministic or stochastic. Dividends can be either payed
at discrete points in time or can be interpreted as a continuous flow of payments.
Furthermore, the amount paid can, of course, be stochastic itself if it depends on
the firms sources of uncertainty. Looking at real options one has to differentiate
between dividends payable to the holder of the underlying (e.g., cash flows if the
investment option is exercised and the project is realized) and dividends payable
to the holder of the real option (e.g. the cash flow earned by the holder of a project

---

[82]See Davis [66].

with a growth option represents a payment to the option holder). The latter is not considered in conventional option pricing theory and therefore makes real option pricing more complicated.

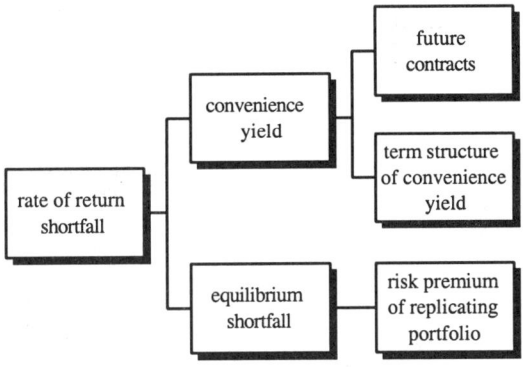

Figure 2.8: Characteristics of the rate of return shortfall

As opposed to financial options, real investment projects often generate cash flows which are not exactly known by time, frequency and amount. This problem is normally solved using a dividend yield or rate of return shortfall that duplicates the dividend payments and explains the below equilibrium return that real investments often earn in comparison to financial assets with otherwise identical risk structure.[83] In case of storable commodities this rate of return shortfall is typically called convenience yield and covers the dividend-like payments that accrues to the owner of the physical commodity but not to the holder of a contract on the commodity.[84] The convenience yield can be determined by future or forward contracts traded at capital markets or by estimating the term structure of convenience yields in case of models with stochastic convenience yield.[85] The rate of return shortfall for other sources of uncertainty needs to be determined by the difference of the expected growth rates of the underlying source of uncertainty and the replicating portfolio in equilibrium,[86] which might be extremely difficult to do in practice because there exists no clear methodology to determine the growth rate of a non-traded source of uncertainty.

Another standard input parameter for option pricing models is the exercise price. For real option models the exercise price is often not constant over the whole lifetime of the real option but may be time dependent due to depreciation or

---

[83] See e.g., McDonald/Siegel [158].

[84] See Trigeorgis [225], p. 95 + 127, Pindyck [188]; Sick [206] on empirical results for oil.

[85] See Schwartz [198, 199], Cortazar/Schwartz [61], Hilliard/Reis [94], Miltersen/Schwartz [169], Moel/Tufano [172]; for an equilibrium approach see Routledge/Seppi/Spatt [194].

[86] See Kulatilaka [126].

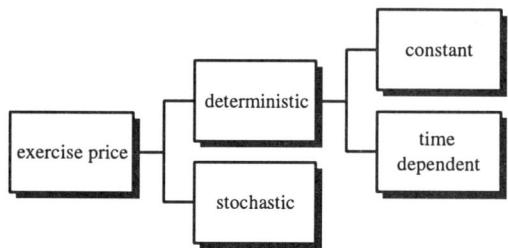

Figure 2.9: Characteristics concerning the exercise price

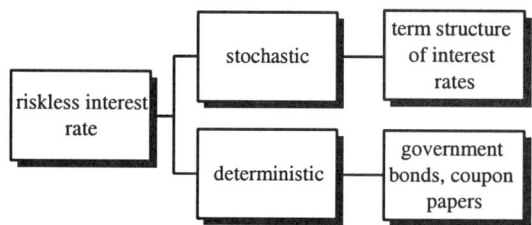

Figure 2.10: Characteristics of the risk free interest rate

contractual restrictions on real investment goods. Furthermore, the exercise price of a real option may be stochastic. For example, if the cost to completion of a product development process is random, then the exercise price of the option to market the new product is random as well.[87] As a consequence, the exercise price itself must be modelled as a stochastic process.[88]

The riskless interest rate is normally assumed to be known and constant for financial option pricing. With real options the assumption of constant riskless interest rates can usually be justified only for small time periods. When it comes to long time horizons the riskless interest rate should be modelled as a stochastic variable itself. Although this has been neglected in most applications of real option analysis because it makes the mathematical models much more complicated, including interest rate risk may lead to striking results.[89] For example, Ingersoll/Ross [104] find that there exists an option value of waiting that is solely due to the stochastic riskless interest rate. Finally, the interest rate can usually be estimated quite precisely using government bonds with zero default risk and the corresponding time to maturity.

The last input parameter for the financial option pricing model is the time to maturity. The time to maturity can either be known or unknown. In the first case,

---

[87] See Pindyck [187] and Weeds [230].

[88] See McDonald/Siegel [160].

[89] See Ingersoll/Ross [104].

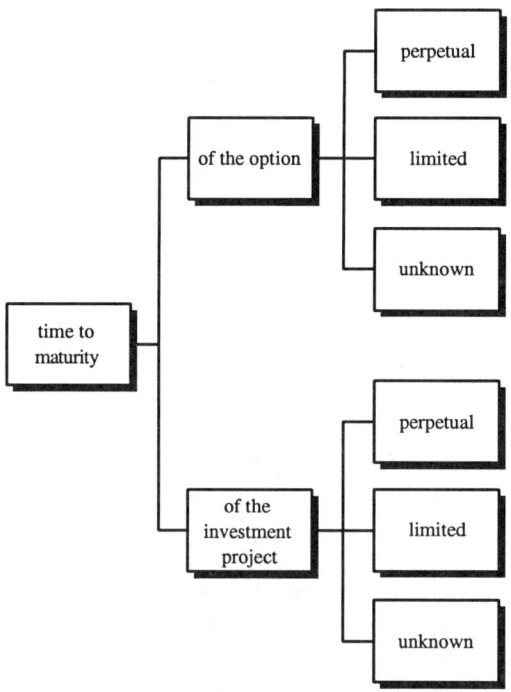

Figure 2.11: Topics concerning time to maturity

one has to distinguish limited or unlimited (perpetual) time to maturity. Because real options are usually not contractually fixed the precise determination of the time to maturity is often not possible. This is especially cumbersome for real option decision situations with a more strategic focus and long time horizons. Furthermore, with real options the underlying itself might expire at some point in the future. For example, a certain product may be substituted by another one making any further growth options written on the demand of the product useless.

Limitations of the Option Analogy

While the above discussed accommodations of the input parameters to the financial option pricing model pose no principle problem for the option analogy to work, there are several limitations that must be treated with special care when modelling real options. These limitations concern the following subjects:[90]

- urgency of decision

- competitive situation

- uncertainty resolvence

- time delay effects

- inter/intra project interactions

Urgency of Decision: The first strategic question of real option pricing refers to the *urgency of the decision*. Management must distinguish between those projects that need an immediate accept/reject decision, that is, *expiring* investment opportunities, and those which management can defer for future action, that is, *deferrable* options. Deferrable projects require a more extensive analysis of the optimal timing of investment, since management must compare the net value of taking the project today with the net value of taking it at all possible future dates. Thus, management must analyze the relative benefits and costs of waiting in association with other strategic considerations, for example, the threat of competitive entry in the meantime.[91]

The next question management has to address in the evaluation process refers to the *exclusiveness of option ownership* and the *effect of competition* on the firm's ability to fully appropriate the option value for itself. If the firm can exercise its real options without being affected by competitors' actions, then the corresponding real option is *proprietary* or *exclusive* to the firm. In this case, the flexibility embedded in the investment project is so specific that competition is precluded from taking effect, for example if the firm is a monopolist or its market position is protected by patents and other strategic insurances.[92] On the other hand, if other firms share the right to exercise the real option and can take away part or all of the project value, the option is *shared* among the firm and its competitors. Therefore, competition reduces the firm's real option value because of the threat of losing part of the project value by competitive actions.

In contrast to many real options, financial options are contracts that provide their owner with an exclusive right to exercise. The holder of the option is not exposed to the danger that other market participants preempt his actions. Exclusive real options may arise also if the company is protected against competitors' actions. This is the case whenever market entry barriers can be created through patents or special knowledge which at least last for some time.

---

[90]These subjects are the basis for the real options classification below. The first three topics were originally raised by Trigeorgis [220], p. 183, resting on results of Kester [110], and were later slightly modified by Meise [163], p. 95. In this study the dimensions *uncertainty resolvence* and *time delay effects* are added.

[91]See Trigeorgis [220], p. 185.

[92]See Pakes [183], Reiss [190], Lambrecht [137].

Competitive Situation: Another limitation of the financial options analogy arises from the fact that real options, unlike financial options, are usually not traded at capital markets. An investor can sell a financial option for its time value. In contrast, the owner of a real option is only able to realize the time value of the incorporated flexibility if the real option can be disposed of, i.e., it is both exclusive and separable from the underlying asset. An example is a patent for a new product or technology. However in general both conditions are not met. The option to expand a producing plant, for example, is not saleable without selling the plant as well. If on the other hand the opportunity to invest is open to other competitors, the value of flexibility is only valuable for its current owner but not for other market participants. In order to avoid anticipated losses of the firm resulting from competitive entry, the only protection is in preemptive exercise of the real option. As a result the option's value may be partially or even completely lost. It becomes clear that real options usually do not represent assets themselves in the sense of financial or real assets which can be traded and sold. They are rather a means for a valuation concept that helps to make superior investment decisions.

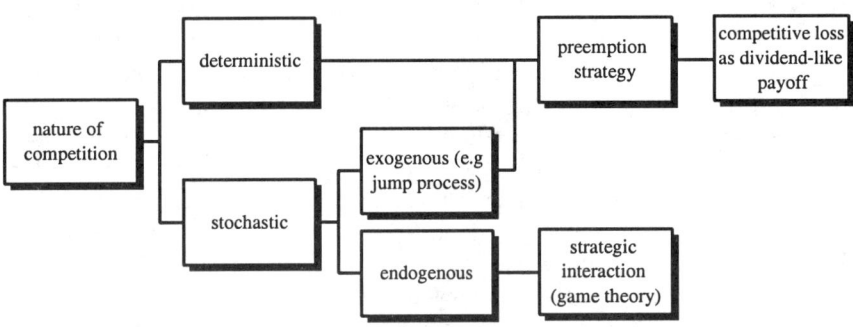

Figure 2.12: Different aspects of competition

From a modelling point of view competitive actions can be considered as either deterministic or stochastic.[93] In case of stochastic competitive actions it can be further distinguished between exogenous and endogenous modelling of competition. Competition is modelled as an exogenous event if the firm has no means to influence other competitors' actions, which is most realistically the case for perfectly competitive markets with many market participants. This kind of competitive assessment is usually modelled as a jump process where only the expected time and impact of competition is taken into account.[94] While the firm usually cannot dispose of the real option that is threatened by competitive actions, the firm's optimal

---

[93] See Trigeorgis [225], Chapter 9.
[94] See Trigeorgis [221] and Dixit/Pindyck [72], section 5.5.B.

reaction is to exercise the real option sooner, trying to preempt other competitors. In the financial option analogy this transforms the consideration of competition to a dividend into like payoff that takes into account the expected competitive losses. If, on the other hand, the market structure is rather oligopolistic, the firm's actions most likely affect the other firms exercise strategies. The resulting strategic interdependencies endogenize the assessment of competition, and it cannot be modelled as an exogenous stochastic event. In order to analyze the strategic interactions among firms, dynamic game theoretic models can be applied to solve the resulting *option games*.[95]

Uncertainty Resolvence: The next important limitation of the financial option analogy is in the kind of uncertainty resolvence of the real option's underlying. As already discussed not all sources of risk are market priced in real option analysis. The exogenous uncertainty of competitive entry might be a good example for private risk that is usually not assigned a risk premium in capital markets.[96] This leads to the question of the investment uncertainty relationship in real option analysis. While the NPV approach suggests that higher uncertainty decreases the project value, this result is not obtained by standard real option analysis. To the contrary, as for financial options an increase in market priced risk *increases* the value of the firm's real options. This intriguing result can be explained by the option asymmetry. While not being exposed to the increased downside risk, the increased upside potential can be exploited by managerial flexibility leading to higher option and firm values. Although this result is often cited as the most important difference between real option analysis and the NPV approach, this is only part of the truth. The favorable aspect of increasing uncertainty is only observable for market priced risk. Private risk, e.g., technological risk, which is solely borne by the firm, *decreases* option and firm value.[97] This highlights the importance of being able to identify and separate private and market priced risk in real option models.

In addition, there is another dimension more directly concerned with uncertainty resolvence. For each underlying source of risk it has to checked how the resolution of uncertainty behaves over time. For financial options both the holder of the underlying as well as the holder of the option on the underlying learn about the underlying's price as time goes by. This is generally the case for all real options exposed to market priced risk. However, there frequently occur situations within real option analysis where the holder of the option does either not learn about the

---

[95]Option games have become an active field of research in recent years in an attempt to capture the strategic components of the real option approach. Among others the most important publications from the perspective of the individual firm are Trigeorgis [220], Smit/Ankum [208], Grenadier [83], Ritter/Haubrich [192], Lambrecht/Perraudin [138, 135], Kulatilaka/Perotti [130, 131], Lambrecht [136, 137] and Weeds [231] and from an industry equilibrium view Baldwin [10], Leahy [141], Caballero/Pindyck [38], Baldursson [9], Chan-Lau/Clark [46].

[96]See the discussion in Section 4.1.2 below.

[97]See e.g., Huchzermeier/Loch [101] for an example of the negative effect of private risk in R&D investments. See also the example on exogenous competition in the next chapter.

uncertainty resolution of the underlying or needs to invest in order to resolve some of the remaining uncertainty. An example for the former is the option to invest in an R&D project with uncertain time to innovation, where the firm does not obtain any further information until it actually starts the R&D project. The latter situation may occur with natural resource investments where the firm must invest in another exploration phase in order to reduce the uncertainty about the well size. These situations are usually referred to in the literature as *learning options*.[98]

Learning options possess two distinctive properties that distinguish them from other real options. First, learning options are usually concerned with the resolution of private risk. Since private risk generally decreases option and firm values, learning options represent an important strategic risk management tool to limit the firm's downside losses and are therefore defensive in nature. Second, from a technical point of view learning options deal with the subject of endogenous sources of uncertainty, i.e., situations in which the firm has the option to influence the underlying sources of uncertainty themselves rather than adjusting its strategy to exogenously given uncertainty. A typical example is marketing expenditure. By placing ads the firm seeks to influence the demand endogenously and therefore tries to directly alter the evolution of the underlying sources of uncertainty.[99]

Time Delay Effects: Financial options as well as most real options that are modelled in the literature assume that the decision to exercise the option takes effect immediately. While this is true for financial options, real life decision making situations often show *time lags* between the decision to invest and the time when the investment is finally implemented and starts generating cash flows. This time delay can be caused by the time it takes to build a production facility, the training of employees, or the development time of a new product to name a few.

Keeping the option analogy, time lags occur either if the underlying asset is received or if the exercise price is payable a certain time period after the decision to exercise the option is made.[100] While the latter causes no principle problem as long as the exercise price is deterministic due to discounting at the riskless interest rate, the former can have a severe impact on the valuation and optimal exercise decision of the real option admitting the time lag. For example consider a firm that faces the decision to invest in a new power plant which takes several years to build. The option to invest in the project is influenced by the uncertainty during the

---

[98] See Copeland/Keenan [59] and Amram/Kulatilaka [7], p. 157, for the general idea of learning options and Huchzermeier/Loch[101] and Kort [122] for the technical treatment within a real option framework.

[99] Since building the capital market analogy for such learning options is extremely difficult, the question of valuing them has not been addressed in the literature so far. Although we will not deal with this subject in the course of the book because we would have to go deeply into the theory of incomplete capital markets, note that the general stochastic control framework developed in the next chapter is capable of dealing with this *direct impact* case. Direct impact refers to management's actions that endogenously alter the underlying stochastic process.

[100] The mathematical details of valuing real options with time lags are presented in the next chapter as one of the examples for the stochastic control framework proposed in this book.

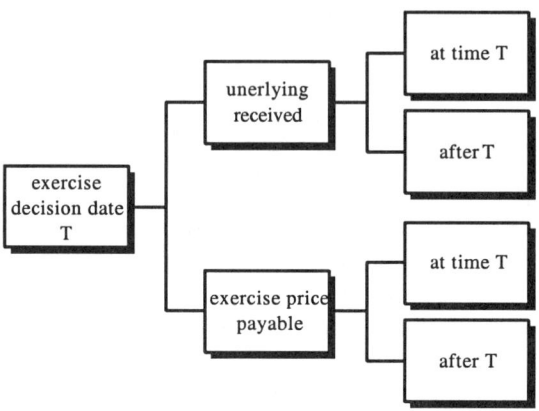

Figure 2.13: Characteristics of time lags

time lag because the firm has to make a commitment now but has to wait until the end of the construction stage in order to learn about the project value it receives in exchange for the investment outlay made today.[101] As a consequence the firm should be more willing to break its initial investment into several stages than in the case where exercising the real option takes effect instantaneously.[102]

Inter/Intra Project Interactions: Finally, the last distinctive feature of real option pricing is the occurrence of more than one real option in the same or across investment projects. If more than one real option is embedded into an investment project, like the option to abandon in addition to the option to defer, we have to analyze the *intraproject compoundness* of the different associated real options. Since the values of real options are typically not additive, it is not feasible to calculate the value of each option alone. We rather have to construct real option interaction models that take account the interplay between the different real options and value them simultaneously.[103] Furthermore, dependencies on other projects have to be recognized as well. This *interproject compoundness* can be differentiated into either *direct dependencies* or *indirect influence on project values*.[104] The former refers to projects which either preclude each other or can only be implemented in parallel or sequentially.[105] Furthermore, projects can indirectly influence each

---

[101] See Bar-Ilan/Strange [13] for a similar example in real estate, Pennings [184], p. 12, for an example of R&D investment with time lags, or Grenadier [84] for an industry equilibrium approach with time-to-build.

[102] See Bar-Ilan/Strange [14].

[103] We will elaborate extensively on this point in the next chapter.

[104] See Kilka [112], p. 123.

others value via their cash flows[106] if there exist, for example, synergies between the projects.

For the above mentioned reasons it is obvious that different firms may assign different values to the same investment opportunity. From a valuation perspective this implies that real option models are heavily context specific and have to be adjusted to each firm's individual circumstances. Moreover, the presence of intraproject interactions calls for the utilization of more complex models than financial option pricing theory is able to provide. While isolated real options can usually be valued using financial option pricing theory with only some minor adjustments, real option interactions demand much more sophisticated models and are not accessible through standard models and valuation procedures. As will be shown, however, in the subsequent chapter stochastic control methods can help to build the required complex option interaction models that allow for modelling and valuing real life investment projects with a large portfolio of associated real options. It will turn out that neglecting some managerial flexibility in an effort to deal with the trade-off between realistically modelling the decision situation and obtaining a parsimonious model may not result in serious valuation errors because the marginal value introduced by additional flexibility is usually quite small.

2.3 Discussion of the Real Options Approach

All the above mentioned special features and limitations of the option analogy make real option valuation more difficult than standard financial option pricing. Although the real option approach is, in principle, applicable to provide a rationale for superior investment decisions in the presence of uncertainty and irreversibility, the crucial point in real option pricing is the art of transferring models developed for financial markets to actual investment decisions. The input required for applying option pricing techniques accurately is, however, often underestimated.

For a rigorous assessment of the value embedded in real options, a detailed analysis of basic variables is required. Only on this firm level can the question of transferability of option pricing models be answered. In order to model real life investment projects with its embedded managerial flexibility, the next question that has to be answered is whether standard option pricing methods are sufficient or whether more complex and sophisticated models should be used. In most cases standard option pricing techniques will yield only a first rough estimation of the true value of flexibility associated with the project, taking into account strategic considerations like competition, time lags and multiple option interactions.

---

[105]See Sanchez [196], Childs [51] and Childs/Ott/Triantis [50], Grenadier/Weiss [85] for examples of each of the above mentioned cases.

[106]See Kilka [112], p. 125 for an example.

2.3.1   When to Use Real Options

Although many real life investment decisions are characterized by uncertainty and managerial flexibility, not all industries are exposed to uncertainty to the same extent. For example, Germany's electric utility industry was characterized until 1998 by serious political regulations that made the firms' business places quite stable. In such situations the use of real option pricing is limited because the standard NPV approach already yields good valuation results. Since the real option approach is rather an extension of the NPV rule than a substitute, this result can be summarized in Figure 2.14.[107]

Uncertainty

		low	high
Room for managerial flexibility	high	Moderate flexibility value	High flexibility value
	low	Low flexibility value	Moderate flexibility value

Figure 2.14: When managerial flexibility is valuable

However, real options should also be taken into account in low uncertainty industries if there exist either significant growth options (as e.g., in R&D projects). Furthermore, managerial flexibility is especially valuable in situations where the NPV approach yields values close to zero because the flexibility to change course is more likely to be used and therefore more valuable.[108]

2.3.2   Advantages of the Real Options Approach

As we have seen, option based models may be used to value investment opportunities. When future market conditions are uncertain but are taken into account in the analysis, the decision to be made is at least partially irreversible, and management has the possibility to shape the implementation of the project. That is, they may be used to model and value investment opportunities having real options and to derive at least initial optimal investment strategies. The option-based approach to modelling real options is conceptually *ex ante* as managers are not committed to undertaking these real options. In reality, capital investments are determined by

---

[107]Copeland/Keenan [58], p. 46.

[108]See Copeland/Keenan [58], p. 46; similarly Amram/Kulatilaka [7], p. 24.

managerial discretion where the available options to invest in real assets are evaluated on an ongoing basis and either, exercised, deferred, or allowed to expire. An option-based approach is therefore a reasonable representation of a managerial decision.

Options are contingent claims and are valued relative to the price of the underlying asset. There are several option pricing techniques that are used for modelling and valuing investment opportunities.[109] Since with all of these methods there is no need to predict future values of the underlying asset, it is sufficient to assume future values to follow a given well-defined process that is stochastic in nature and represents the uncertain fluctuations of the underlying asset. Specifically, uncertainty over the value of the underlying asset is usually expressed in terms of the volatility σ. The beauty of the option pricing techniques lies in the possibility of applying the risk free rate, versus a risk adjusted discount rate when using DCF or an ambiguous discount rate in the case of decision tree analysis, for valuation purposes. This greatly simplifies the valuation procedure for real options since the methodological connection to capital markets places the real option approach on solid theoretical grounds. It therefore avoids the issues of risk preferences and risky discount rates by using the risk free rate and risk neutral (equivalent) probabilities.[110] Furthermore, the real option approach eliminates having to estimate the expected rate of change in the underlying asset.

In comparison to DCF the modelling of uncertainty results directly from the volatility σ versus indirectly via a risk adjusted discount rate. Together with the observed asymmetry of the distribution of investment opportunity values, real option theory provides a means to derive the quantitative impact of managerial flexibility. It is therefore a very flexible and potentially simple yet powerful decision-making framework. In its ability to account for, and quantify, active project management, time dependencies, project interactions and interdependencies, and option interactions, it is specifically designed to model flexibility and delivers consistent resulting valuations.[111] Furthermore it can, at least conceptually, be used to model and value many types of business decisions and, in some cases, easily model and value reasonably complex investment opportunities.

Another distinctive feature of real option pricing is that it not only gives a rational value of the investment opportunity, but also determines optimal strategies to manage the project. In case of the option to defer, for example, it tells management when and under what conditions it should invest in the project. Even if the problem formulation is very complex, lacks relevant data, or even violates assumptions un-

---

[109] See the following chapter.

[110] The term "risk neutral" is somewhat messy in that context since, in fact, the results of option pricing theory are independent of any investors' preferences. It just states that due to the use of the risk free rate discounting may be performed *as if* all investors were risk neutral. Hence, we will rather use the term "equivalent probability measure" in the sequel.

[111] Consistent is meant from a capital market point of view. The real option approach allows for risk averse investors without having to specify the degree of individual risk aversion explicitly. It is consistent with an equilibrium price structure and produces therefore *rational* valuation results.

derlying the option-approach, it still provides management with valuable insights and helps to put a structure on the decision making process.

In addition, the real option approach may explain the investment uncertainty relationship. Generally, an increase in the variance leads to an increase in the range of possible future values of the underlying asset. Yet, due to option asymmetry, the maximum possible gain increases while the maximum possible loss does not change. Therefore, option values generally increase as market priced risk increases.

An option-based model is also highly appropriate when there are sequential, or phased, projects, since the flexibility of multistage decision making is explicitly taken into account. This issue is especially important for the valuation and justification of projects that do not yield any profits immediately but are of great strategic importance to the firm. This may be the case for R&D projects or other types of strategic investments that can be characterized by compound options.

Lastly, the ability to use spot and futures prices for several sources of uncertainty makes an option pricing model a powerful and robust analysis, as traded securities are then available for parameter estimation and the capital market analogy works properly. This also means that subjective probabilities are not needed. Moreover, future values and cash flows are market determined, circumventing the need to project and estimate future cash flows.

In summary, the real option approach has not only the potential to overcome the obstacles of standard valuation methods but, more importantly, allows one to quantify strategic issues of firms from a capital market point of view leading to disciplined corporate strategies. It therefore has the potential to close the gap between capital market theory and strategic management.

2.3.3   Drawbacks of the Real Options Approach

It is sometimes argued that real option pricing contradicts fundamental insights of classical capital budgeting research. This criticism is based on the comparative statics of option pricing theory which produce contradictory results to former methods, especially DCF. First of all, more uncertainty in general raises the value of a project while in contrast higher uncertainty was believed to reduce value because of the higher risk premium investors would demand for riskier projects. Furthermore, higher risk free interest rates raise option values, too, which states again a fundamental difference from DCF-based methods which are based on the time value of money. Lastly, critics note that the value of an option increases with the time to expiration, although an investment was intuitively said to have less worth the longer the positive cash flows are in the future.

From a scientific point of view it is not appropriate to criticize new theories because of their opposing results to old ones. On the contrary real option pricing was designed to overcome weaknesses or even mistakes of traditional concepts in capital budgeting.[112] As long as real option pricing is able to explain real world

---

[112]See Meise [163], p. 83.

phenomena that traditional capital budgeting methods are not capable of, it is certainly preferable to traditional concepts. Especially, if the new method is able to recognize up to now neglected parts of project value, namely the value of associated real options.

Option-based models are not, however, without their weaknesses. Several disadvantages are mentioned in the literature. Criticism may arise from the following fields:

- capital market assumptions

- modelling and valuation process

Capital Market Assumptions

In order to value real options as contingent claims using financial option pricing theory, the conceptual analogy between financial and real options is certainly not sufficient. To the contrary, one has to make certain that the assumptions underlying the valuation of financial options are fulfilled for real options.

These assumptions of financial option pricing theory as a capital market oriented preference free valuation methodology consist especially of the perfect and complete capital markets hypothesis. Especially, the assumption concerning market completeness plays an important role in the valuation of real options because it ensures that the spanning condition is fulfilled. The spanning condition requires that the option's underlying itself or a perfectly correlated portfolio of assets is traded in the market such that the stochastic component of the underlying can be duplicated pathwise. This condition guarantees that the fair market price of the real option can be found even if the real option's underlying is not traded.

Another possible criticism that is sometimes mentioned in the literature[113] is that the real option's payoff cannot be replicated by a trading strategy since taking a short position in the investment project is impossible due to the fact that the project itself is not traded. However, this criticism is misplaced and fails. The construction of a replicating portfolio, as proposed by Black/Scholes [27] and Merton [165], is not the only way to price an option. Since the important results of Harrison/Kreps [88] and Harrison/Pliska [89] it is well known that the determination of an *equivalent martingale measure* yields rational option prices as well without requiring liquidity of the underlying asset. By way of contrast, their result implies that not only options on nontraded assets but also options on *nonexisting* assets may be priced using option pricing theory.[114] This property is especially useful in the context of real options where the underlying asset — remember that the underlying asset in many real option applications is either the project value, a pricing factor or cash flows — may even not yet exist, as for example is the case for the option to wait to realize a project.

---

[113] See Amram/Kulatilaka [7], p. 57.
[114] See Sick [205].

To conclude, the applicability of real option pricing is not restricted by the lack of trading in the real options themselves or the underlying assets. It has rather to be checked in every specific case whether one can fall back on a portfolio of market priced assets with the same risk structure as the real option's underlying. Conversely, every real option that is written on a market priced asset, which can be found for example for mining and many natural resources projects, can be valued without any difficulties using the real option approach. Nevertheless, there exist many projects, like R&D of a new product, which are not contained in the existing investment opportunity set and spanning might therefore fail to apply.

If markets are incomplete, i.e., there exist sources of risk which cannot be spanned by capital markets, the real option approach has to fall back on individual investors' risk preferences towards these sources of risk again exposing the valuation results to subjective judgements. However, this is not only a problem of the real option approach but of all other capital budgeting methods as well. The results obtained by a real option analysis in incomplete markets can at least be interpreted as the lower limit on an overall option value, which informs the principle judgement of the favorableness of an investment project.

Market incompleteness is especially cumbersome in that situation where it most frequently will occur. That is, in long term strategic investments where the option value is by far the highest part of the firm's expanded net present value. Unfortunately, there is no easy way out. Despite recent advances in the real option theory of incomplete markets,[115] there is no solution to this problem available to date. However, even with this serious obstacle in applying real option thinking for valuation purposes, it still disciplines managerial decision making and can serve as a qualitative guideline in strategic project management.

Modelling and Valuation Process

Problems with modelling include the identification of relevant sources of uncertainty, modelling the appropriate stochastic processes, spanning these uncertainties, and estimating the corresponding parameter values. Furthermore, all possible and relevant real options inherent in the investment opportunity might be difficult to identify, which subsequently introduces uncertainty over the firm's strategy space.

Moreover, from a modelling point of view the tractability of investment opportunities that are characterized by many real options is still limited due to the lack of appropriate valuation techniques. Option interactions can cause very complex models and valuation procedures which are not well understood. And while most real world problems usually contain more than just one single real option, the valuation process of such complex models almost never yields closed form solutions for project values. Consequently, one has to rely on numerical methods which in turn are hard to communicate to senior management since they cannot be understood without some in-depth knowledge about option valuation techniques.

---

[115]See especially Lee [142].

2.4 Conclusions

In this chapter we demonstrated why and how the real option approach overcomes the disadvantages of traditional capital budgeting methods in that it takes into account flexibility, uncertainty and irreversibility in investment decisions. It provides a capital market rational that allows one to link the firm's strategic decisions with its objective of maximizing market value. We have described how the bridge to capital market valuation can be built using the analogy of real and financial options. We argued that the objective function of the firm must be extended to take account of the real option value of managerial flexibility. Furthermore, the existing research done in the field was mentioned by taking a management and a valuation perspective of real options which allowed us to define and clarify the main nomenclature and to describe the different difficulties associated with applying real option valuation. Finally, the advantages and disadvantages of the real option approach were discussed.

Although the mathematical models usually employed to value real options were not presented, we conclude that one of the most serious obstacles for a faster and further adoption of the real options approach among researchers and practitioners is the lack of financial option pricing techniques to deal with all the special features associated with real life investment projects. Especially, its inability to deal with complex real option interaction models and the limitations discussed in Section 2.2.2 call for other approaches that are capable of dealing with complex real option models without dropping the link to capital markets. In the following chapter such an approach is proposed by connecting the theory of mathematical finance with impulse control and optimal stopping problems of stochastic control in order to develop a stochastic control framework for valuing complex real option interaction models.

Chapter 3

Real Options and Stochastic Control

3.1 Real Option Interactions and Stochastic Control

The aim of this chapter is to derive a modelling framework for valuing projects with multiple incorporated real options. The motivation of such efforts is clear. A decision maker is usually not interested in just a value of a single option but rather in the influence of a certain option on the value of the whole project. But since option values are in most cases not additive, one has to explicitly take account of the contingency structure of the incorporated real options in order to find the fair value of the project.[1] It is therefore important to find a framework which allows in a stringent way to value projects in the presence of option interactions. Furthermore, such a modelling framework should have the necessary generality to deal with many different problems where sometimes completely different options are present.

So far there has not been developed such a general framework. Most work in this field was concerned with special cases and examples. Brennan and Schwartz [34] were the first to develop a model with several embedded real options in order to value natural resource investments. On a more microeconomic level, Dixit and Pindyck analyzed entry and exit decisions using option pricing theory under various settings in order to explain the so-called hysteresis effect of the investment behavior of firms.[2] Several authors extended their work considering uncertain cost,[3] uncer-

---

[1] See e.g., Trigeorgis [223, 224].

[2] See their book [72] and the references therein for an excellent review of real options applied to microeconomic theory of investment under uncertainty and industry equilibrium.

[3] See Pindyck [187].

tain learning effects,[4] optimal capacity choice,[5] and the flexibility to temporarily shut down and reopen production.[6]

Trigeorgis [223, 224] built up a hypothetical project with the flexibility to wait, the flexibility to abandon after each building step during the construction phase, the option to expand at some points in time and the flexibility to abandon operation. He used a discrete time setting to value these options with binomial (trinomial) trees. Since he did not give a thorough formal exposition of his model, his methodology seems somewhat unclear but he provides some very interesting and intuitively appealing results.[7]

A first step towards an integrated model that tries to take account of different real options in one project was given by Kulatilaka and Marcus [127] and was later extended by Kulatilaka [123, 124, 126, 125]. The basic idea in Kulatilaka´s papers is to model flexibility as the ability to change between different modes of operation. That means the firm has the option to change from one state to another by paying a predetermined switching cost, i.e., the exercise price in terms of option pricing theory. States or operation modes — as Kulatilaka calls it — are for example currently producing product A vs. producing product B or invest now vs. invest later. Therefore, Kulatilaka developed a discrete time switching model using stochastic dynamic programming to value a project with different embedded options. For example, he considered a project containing the option to defer, the option to shut down production, and an additional growth (expansion) option. Like Trigeorgis, he derives the result that the incremental value of any further flexibility option introduced into the project is neither additive nor is its effect in comparison to other real options independent of the level of uncertainty expressed through volatility. Since his model is discrete in time and states, his generalized switching option model consists of many nested European type compound exchange options.

A continuous time approach for a generalized switching option was developed by Triantis and Hodder [219, 96] for valuing flexible manufacturing systems. Their model provides a formal derivation of how to use contingent claims analysis to value projects which involve several states of operation and exogenous uncertainties. Their work was so far the most general that has been done to take account of valuation of projects including option interactions.

Unfortunately, there exist some shortcomings of the above mentioned models. First of all from a technical point of view they either use only American — usually perpetual — or European type options exclusively, which is somewhat unrealistic when considering the complexity and time dependence of real life projects. Secondly, the influence of different combinations of real options on project value is studied only qualitatively. It is quite clear that two options to change from state A to state B at different but close points in time are not worth much more than only

---

[4]See Majd and Pindyck [151].

[5]See Pindyck [186], He and Pindyck [93].

[6]See Dixit [71].

[7]While his example was so far the most interesting concerning option interactions, we will take up his model later in this chapter in order to demonstrate how it fits into our framework.

one of these options alone, since both have the same direction and the probability that both would be exercised is usually quite large. Alternatively, two options with different switch directions can be almost additive in value because they cover more of the state space. Therefore, the joint exercise probability is small in comparison to the former case. The third problem is that exercise of an option may alter the value of the underlying asset itself and thus will change the value of all subsequent options. This kind of compoundness requires a detailed analysis of the contingency structure of the incorporated real options.

Briefly, the degree of interaction between real options depends on:[8]

- whether they are written on the same underlying,

- whether the options are of the same type (direction), i.e., put or call options,

- the separation of their exercise time (also considering American or European exercise regions),

- their relative degree of being in or out of the money, and

- their order or sequence.

Numerical results suggest that there exists a maximum of flexibility where the marginal cost of acquiring an additional real option — in the sense of building it into the project — exceeds its marginal contribution to the investment value.[9] But it is still unclear which portfolio of real options management should incorporate or at least which rules it should follow. Furthermore, most models concentrate just on timing options, not allowing for control variables that have to be chosen optimally by management, like implementation options, e.g., the choice of capacity when initiating a production facility.

So far a theory of option interactions in order to get quantitative results to determine which combination of flexibility options to implement into a project is still missing. Establishing such a theory seems to be crucial to achieve a broader acceptance of real option pricing because it provides us with the possibility to value more realistic investment opportunities in practice rather than valuing just idealized projects with little flexibility. Such a general framework should satisfy certain requirements. It should

- be flexible and general,

- be easy to implement and adjust,

- consider different types of options,

- be understandable for management,

---

[8]See Trigeorgis [220, 223, 225].
[9]See Kulatilaka [125].

- prevent overmodelling,

- build bridges to capital markets,

- give not only the value of an investment but strategic rules as well.

As for the valuation of financial options, generally a continuous time formulation of real option problems can be found in the literature, allowing for the use of *contingent claims analysis* as developed in the literature on pricing financial options.[10]

For financial markets, the continuous time formulation is reasonable since information arrival, trading, and price movements occur almost continuously in active markets. Moreover, the continuous time framework provides convenient analytical properties for pricing options and, for some problems, yields closed form solutions. For less tractable problems, numerical procedures have been used to identify approximate solutions to an underlying continuous model.

With regard to both the literature and practice of capital budgeting, continuous time models have not been widely utilized. However, for many capital budgeting applications, a continuous time approach is at least as natural as the annual discrete time formulations commonly employed. Furthermore, it allows for using results and techniques developed in the financial options literature to value real options. Consequently, a continuous time approach is taken throughout the chapter.

It will prove useful to view project control opportunities (managerial flexibility) as the ability to change the *state* of the project. These states may be discrete (e.g., an extracting or closed oil field,[11] developed or undeveloped land[12] etc.) or continuous (e.g. operating at different levels of capacity[13] etc.). The real options literature contains various examples of options to change the state of a project. For example, the option to initiate a project[14] or the option to abandon[15] both involve an irreversible decision which precludes switching back to the previous state. In contrast, the option to temporarily shut down[16] or the option to switch inputs and outputs[17] are reversible decisions.

Therefore, two qualitatively different kinds of switching that occur in real option models can be distinguished. The first is where switching between alternative states is (costly) *reversible*, and the second where changing the state of the project is *irreversible*. The latter case exhibits the usual features of financial options, since the holder of a financial option has usually the right to exercise the option only once. In

---

[10]See e.g., the early contributions by Brennan/Schwartz [34] and McDonald/Siegel [159, 160] for real option problems formulated in continuous time resting on the results of financial option pricing in continuous time as summarized by Merton [166].

[11]See the extensive real options literature on applications in the natural resources industry.

[12]See e.g., Grenadier [83], Quigg [189], Titman [217], Capozza/Li [42].

[13]See e.g., Bollen [28], the case study later in the chapter.

[14]See McDonald/Siegel [160].

[15]See Myers/Majd [178].

[16]See McDonald/Siegel [159] or Dixit [69].

[17]See Margrabe [153].

contrast, many managerial decision problems contain situations where subsequent switching between different states is possible, exhibiting a complex compound option structure with many nested options, which complicates the valuation of such projects using classical option pricing theory.

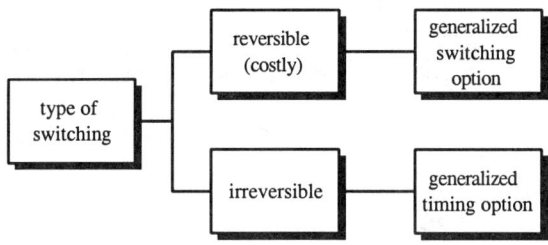

Figure 3.1: Reversible and irreversible switching

A further examination of the characteristics of real options leads to the following differentiation. First, only options that are costly to exercise are examined. Furthermore, keeping the option's analogy, it is assumed that costly switches incur a fixed cost part, and possibly a variable cost part depending on the state of the project, which together form the exercise price of the real option.

Second, for demonstration purposes only American type options are considered in the exposition of the two basic real option models.[18] So that changes in the states of the system may take place in a prespecified time interval during the life of the project or firm.

With regard to the distinction between reversible and irreversible switches, any real option model can be decomposed into combinations of *generalized switching options* where decisions are costly reversible and *generalized timing options* where decisions are irreversible, see Figure 3.1. One may think of the following example. Consider a firm that consists of a plant producing a certain product. The firm may have the option to temporarily shut down and restart production. Furthermore, it may have the option to completely switch capacity of the plant to a new product for which the firm holds an exclusive patent that expires at some time in the future. After exercising the option to switch from the old product to the new one, the firm still holds the option to temporarily shut down and restart production, but may not for economic reasons switch back to the old product. Clearly, the problem of valuing the firm can be decomposed into two generalized switching options — one for manufacturing the old product and one for the new product — which are separated by the timing option to utilizing the patent and changing from the old to the new product which is irreversible.

---

[18]This assumption will be relaxed later in the chapter where the contingency structure is introduced.

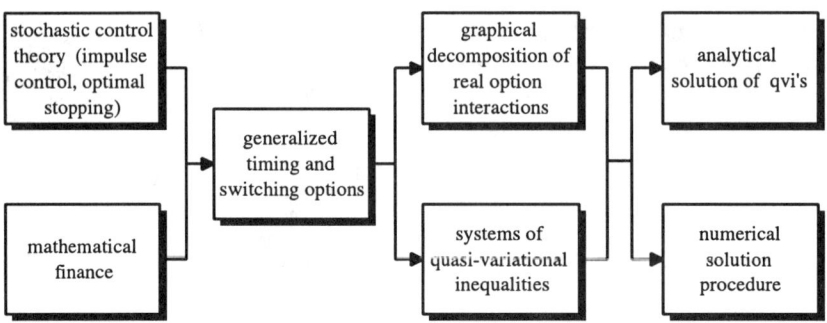

Figure 3.2: Organization of the study

Using the idea of decomposition of real option interactions the chapter is organized as follows. In Section 3.2 it is argued that the theory of stochastic control can
be applied to value option interactions. Specifically, generalized switching options
can be valued using *impulse control theory* while for generalized timing options
the *theory of optimal stopping* in a generalized sense can be employed. We will
start by giving a mathematical introduction into impulse control theory, taking also
into account the considered stochastic processes used to model the firm's environment. Afterwards general impulse control and optimal stopping models specifically
designed to value real options are introduced. Furthermore, proofs of verification
theorems´ for both cases are given, which put the results on mathematically solid
grounds.

The next step is to establish the link to capital markets in Chapter 4 where methods of mathematical finance are used to find the market value of the firm which we
are primarily interested in. It turns out that the market values of generalized timing
and switching options are the solutions of systems of *quasi-variational inequalities*
derived from the equivalent optimal stopping and impulse control problems. After
that a very convenient graphical method to represent the contingency structure of
real option interactions is presented and the suitability of the proposed stochastic
control framework is demonstrated by some examples.

In Chapter 5 various extensions along the line of the above mentioned real
options classification are given. It is shown how they fit into the stochastic control
framework. In addition some new types of real options are mentioned which naturally drop out if one uses the stochastic control framework, also demonstrating its
power as a valuation tool.

In most cases analytical solutions cannot be found and numerical solution techniques have to be employed. The practical dimension of the decomposition method
proposed becomes obvious because the contingency structure of the stochastic con-

trol problems can easily be transformed into numerical valuation schemes using finite difference methods. To demonstrate how this can be done in a specific situation, a case study dealing with the valuation of strategic flexibility in the manufacturing industry is provided in Chapter 6. Its mathematical and numerical treatment is discussed in detail.

Finally, we conclude with the managerial implications of the proposed valuation approach and how it can help to integrate aspects of real option valuation and control into managerial practice.

3.2 Introduction to Impulse Control and Optimal Stopping

3.2.1 General Idea

In this section we will give a short introduction to stochastic optimal control. To serve our purposes of integrating real option problems into a stochastic control approach we will restrict the exposition of the mathematical models to a tractable and understandable level without completely dropping the mathematically necessary thoroughness.

Stochastic control problems can be roughly divided into three generic classes. First, problems where the decision maker has the means to react continuously to the realization of a stochastic process are called *continuous stochastic control problems*.[19] Furthermore, the degree to which the decision maker can choose when he can optimally exit a stochastically evolving system determines the class of *optimal stopping problems*. Lastly, problems where the decision maker intervenes in the system only at specific irregular times and optimizes over certain available decision variables denote the class of *impulse control problems*.

Optimal stopping problems are by nature very similar to the pricing of American options because, in the case of American options, the holder of the option has to determine the optimal exercise time. This corresponds to choosing a stopping time and exiting the system by selling the option. In contrast, impulse control problems allow for much more complex structures. In option pricing terms securities can be valued using impulse control, which allows the holder to switch between a variety of portfolios, each with a different composition of the underlying assets. The exchange between portfolios could take place at prespecified prices, and there may be a cost associated with each exchange. Impulse control models can therefore take two important features of real option pricing into account. First, exchanging

---

[19]This type of control problem will not be pursued here due to its lack of option-like properties. However, the integration of continuous control problems into our general real options framework would not be too difficult. For an interesting example where continuous control and impulse control is combined see Mundaca/Øksendal [175]. See also Kamrad/Lele [108] for an example of combined maintenance and insurance policies using continuous stochastic control and the real options approach.

two or more risky assets against each other and second performing exchanges frequently in time and therefore admitting (costly) reversibility and option compoundness.[20] [21]

The traditional approach to solving optimal control problems is the same for each of the three above mentioned classes. First, intuition and heuristic arguments are employed in order to derive the dynamic programming or Hamilton–Jacobi–Bellman partial differential equations. This corresponds to finding the fundamental pricing equation, for example in a Black/Scholes market model.[22] Second, a verification theorem which relates the solution of the dynamic programming PDEs to the problem's value function is proved under certain appropriate assumptions. Third, the problem of existence and uniqueness of the solution of the PDEs which satisfy the assumptions of the verification theorem is addressed, and numerical schemes are employed, or, in rare cases, explicit solutions are derived. We will follow exactly that procedure for the generalized switching option and the generalized timing option.

3.2.2   Impulse Control

Introduction

The basic idea of impulse control is the observation that managerial decisions under uncertainty do not take place permanently but irregularly. Taking managerial actions usually incurs some fixed cost of intervening into a system and are often costly reversible. For example, the decision to enter a new market incurs fixed cost because of marketing efforts, building new production facilities and supply chains etc. Furthermore, it preserves the ability to later exit again which makes the entry decision at least partially reversible. For those reasons the optimal managerial strategy is to take action whenever certain events occur and perform the optimal action. This implies that management only acts at certain time instances rather than permanently and that after each intervention there exist time periods where the system evolves uncontrolled.

The opportunities for modelling economic problems with use of impulse control techniques seem to be vast and unlimited since virtually any control by management can be characterized by discrete intervention times and some fixed cost part of intervention. Yet, it is very surprising that so few examples of impulse control applications in economics and management can be found in the literature. This may be due to the nontrivial mathematical machinery one has to employ when

---

[20]The option to exchange was first studied by Margrabe [153] where stock prices follow geometric Brownian motion and was later extended by Vollert [229] to the case of Paretian stable asset return distributions resting on previous work by Hurst/Platen/Rachev [103].

[21]Geske [80, 81] developed a model for compound options which was later combined by Carr [43, 44] with the model of Margrabe for the study of a sequential exchange option in the European and American option setting.

[22]See e.g., Trigeorgis [225] for the derivation of a fundamental PDE for valuing real options or one of the numerous books on mathematical finance, e.g., Duffie [75].

using impulse control. Also, analytical solutions for the optimal control policy are rarely found and do not produce as intuitive insights as solution methods for which closed form expressions exist. On the other hand, since computational methods are readily available and computer performance is increasing year by year,[23] the task of implementing numerical schemes for impulse control problems is offset by the appealing results it can produce.

The idea of modelling managerial decisions as impulse control problems goes back to Bensoussan in the early 1970s.[24] Bensoussan/Tapiero [21] provided the first article where possible applications of impulse control were mentioned. Except for some early precursors on optimal cash management by Constantinides/Richard [57] and Constantinides [56], it took several years until impulse control models were applied to other areas of managerial decision making. Harrison/Selke/Taylor [92], Sulem [215, 214] and Bar-Ilan/Sulem [16] used impulse control methods to model inventory problems. More recently, Jeanblanc-Piqué [105] and Mundaca/Øksendal [175] considered optimal interest rate and exchange rate controls by federal banks.

In financial theory the only field where impulse control has been actively applied so far is in portfolio optimization with transaction costs. The models proposed by Eastham/Hastings [76] and Davis/Norman [67] were recently extended by Korn [120, 121]. In the real option literature, impulse control models are rarely found. Although the famous hysteresis model of entry and exit decisions under uncertainty by Dixit/Pindyck [72] is in fact nothing else but the simplest possible impulse control model, it has not been mentioned by the authors. However, there seems to be a growing interest in using impulse control in real option analyses as the articles of Mauer/Triantis [157] on optimal financing and capital structure decisions and of Hodder/Triantis [96] on process flexibility suggest. The most recent application of impulse control dealing with optimal re-engineering policies can be found in Liu/Yao [146].

Mathematical Idea and Considered Stochastic Processes

Mathematical Idea: It will prove useful to start by describing the general mathematical intuition of impulse control models. Consider an arbitrary economic system that is exposed to an uncertain environment modelled as a stochastic process $U(t)$. Management has the means to intervene in this system by changing the state process $U(t)$ to the new state $U(t) + \zeta$ at arbitrary times t. However, taking the control action ζ incurs switching costs $H(t, \zeta) > k > 0$ with a fixed cost component $k > 0$. It is obvious that a strategy to continuously control the stochastic process $U(t)$ would immediately result in bankruptcy since the fixed costs would add up to infinity even in marginally small time intervals.[25] But what is the optimal control policy to apply in this setting? Consider the following strategy. At any time we

---

[23]We discuss the topic of numerical solutions of impulse control problems in Chapter 6.

[24]See the book Bensoussan/Lions [20] and the references therein.

[25]See Korn [121].

have to make the decision whether to apply a control action or not. Assume that
$\Phi(s, u)$ is the optimal objective (value) function at time s with state $U(s) = u$. If
at time $t = s$ it is optimal to apply an instantaneous control ζ, then the state at this
time would jump by ζ. Consequently, the objective function Φ would take on the
value $\Phi(s, u + \zeta)$. Then, at the intervention time we have to take into consideration
the cost to exert the control. Hence the new value of the objective function is given
by $\Phi(s, u + \zeta) - H(s, \zeta)$. While we would not apply an impulse control if the
optimal objective function $\Phi(s, u)$ (recall that Φ results from a maximization) is
greater than the optimal objective function right after an impulse control is applied,
we can state that

$$\Phi(s, u) \geq \sup_{\zeta} \{H(s, \zeta) + \Phi(s, u + \zeta)\} = \mathcal{M}\Phi. \qquad (3.1)$$

However, the inequality holds as an equality when an impulse control is triggered.[26]

If we do not apply an impulse control, then Φ will evolve according to the
stochastic nature of the underlying state process $U(t)$. Let therefore Φ_1 be the
uncontrolled objective function. Obviously, we have at any time s and for any
value u that

$$\Phi(s, u) \geq \Phi_1(s, u), \quad \Phi(s, u) \geq \mathcal{M}\Phi. \qquad (3.2)$$

The intuition behind this result is that during time intervals where it is not optimal to intervene, $\Phi(s, u)$ is equal to $\Phi_1(s, u)$ and $\Phi(s, u)$ is strictly greater than
$\mathcal{M}\Phi$. Otherwise intervention would be optimal. On the other hand, at times where
intervention is triggered the second inequality is an equality. But now the value
function with impulse control $\Phi(s, u)$ must be greater than the value function without control $\Phi_1(s, u)$. Therefore, for both cases one of the above inequalities must
hold with equality while the other inequality is strict. We can summarize the two
inequalities above by the relation

$$[\Phi - M\Phi][\Phi - \Phi_1] = 0. \qquad (3.3)$$

Finally, we add the boundary condition[27]

$$\Phi(T, U_T) = 0. \qquad (3.4)$$

Bensoussan/Lions [20] proved that this set of relations defines the function Φ, if
it exists, in a unique way. In order to obtain a solution $\Phi(s, u)$ we may proceed as
follows. Define a *continuation region* which fulfills the following relation

$$\mathcal{C} := \{(s, u) \,|\, \Phi(s, u) > \mathcal{M}\Phi(s, u)\}. \qquad (3.5)$$

---

[26] $\mathcal{M}$ denotes a maximum operator in this case. However, $\mathcal{M}$ might also denote a minimum operator
if our objective is to minimize cost rather than maximize value.

[27] As we will see later this boundary condition — or terminal value — may also be a function of the
state variable $U(t)$ or even depend on the result of subsequent optimization problems.

The continuation region $\mathcal{C}$ contains any combination (s, u) for which intervention is not optimal and the system evolves uncontrolled. Of course, for the uncontrolled system, $\Phi = \Phi_1$ holds because if the second inequality of (3.2) is strict the first inequality has to be an equality.

Furthermore, define $\theta_1 > s$ as the first time when $(\theta_1, U(\theta_1)) \notin \mathcal{C}$ and $\zeta_1 = \zeta(\theta_1, U(\theta_1))$ as the corresponding optimal impulse control. The function $\zeta(s, u)$ results from

$$\sup_{\zeta} \{H(s, \zeta) + \Phi(s, u + \zeta)\} = H(\zeta(s, u)) + \Phi(s, u + \zeta(s, u)). \qquad (3.6)$$

That is, for each impulse time Θ_i the corresponding impulse control ζ_i is chosen such that the value of the system right after the impulse control is maximized. Equation (3.5) gives the new optimal objective function right after the jump ζ. Thus, we are able to derive the first optimal impulse ζ_1 at the first corresponding optimal impulse time θ_1. Iterating this procedure we may construct step by step the optimal impulse control strategy $\tilde{w} = (\theta_1, \theta_2, \theta_3, \ldots ; \zeta_1, \zeta_2, \zeta_3, \ldots)$ consisting of all optimal intervention times and their corresponding optimal control. Although this solution procedure seems to be quite simple, it is difficult to apply in practice. With certain exceptional cases, impulse control models will not yield closed form solutions for the objective function Φ. This is mainly due to the fact that impulse control problems consist of sequential optimal stopping problems with additional embedded control variables for which optimization has to be carried out simultaneously.

In the following we will discuss impulse control models in a mathematically more rigorous way, also taking into account the specific type of fundamental stochastic processes that fit our purposes for modelling and valuing real options.

Considered Stochastic Processes: While the above mentioned results are generally independent of the specific type of stochastic behavior of the system, we will only consider in what follows stochastic processes that consist of Brownian motion and Poisson jump components. Under this assumption we will reformulate the impulse control problem accordingly.

Assume that we are given a dynamic system which is completely characterized by its state. At time t, the state of the system is described by a vector $U(t) \in \mathbb{R}^n$. This implies that all relevant quantitative variables are assumed to be real.[28] Furthermore, the process $U(t)$ is a stochastic process whose evolution is modelled by a diffusion with jumps.[29]

This stochastic state vector is subject to the usual filtered probability space $(\Omega, F, \mathcal{F}, \mathcal{P})$ for times $t \geq s$. $\mathcal{F} = \{F_t : t \geq s\}$ denotes an increasing, right

---

[28] In this framework we are able to handle state variables that are integers as well.

[29] An introduction to stochastic processes in continuous time and especially to Brownian motion can be found in Harrison [87]. Introductions with a more finance related focus can be found in Shimko [203], Lamberton/Lapeyre [134], Shreve [204] and Karatzas/Shreve [109] in ascending order of mathematical rigor.

continuous, and augmented filtration with respect to the above probability space. Let $B = (B^1, B^2, \ldots, B^d)$ be a d-dimensional standard Brownian motion on $\mathbb{R}^d$ that is a martingale on the filtered probability space $(\Omega, F, \mathcal{P})$. Moreover, let

$$s \leq \tau_1 \leq \tau_2 \leq \cdots \leq \tau_i \leq \cdots \tag{3.7}$$

be the random points in time when the Poisson jumps occur.[30] Then we have

$$
\begin{aligned}
dU_t &= \mu(t, U_t)\, dt + \sigma(t, U_t)\, dB_t, \qquad t \in [\tau_i, \tau_{i+1}), \\
U_{\tau_{i+1}} &= U_{\tau_{i+1}^-} + \xi_{i+1}, \\
U_s &= u,
\end{aligned}
\tag{3.8}
$$

where ξ_i denotes the size of the jump at time τ_i. In the interval $[\tau_i, \tau_{i+1})$, the process U_t evolves according to a continuous diffusion with drift vector $\mu(t, U_t) \in \mathbb{R}^n$ and diffusion matrix $\sigma(t, U_t) \in \mathbb{R}^n \times \mathbb{R}^d$ which are non-random and fulfill Lipschitz and growth conditions. Equivalently, this can be stated in the differential equation form

$$
\begin{aligned}
dU_t &= \mu(t, U_t)\, dt + \sigma(t, U_t)\, dB_t + \sum_i \xi_i\, \mathbb{I}_{\{t \geq \tau_i\}}. \\
U_s &= u.
\end{aligned}
\tag{3.9}
$$

Note that the sizes of the jumps ξ_i at points τ_i where the event of a jump occurs may itself be a random variable depending on the current state of the system (τ_i, U_{τ_i}). Although this is a quite general notation of a Poisson jump process, it is assumed for simplicity that the distribution of the jumps can be decomposed into a deterministic function $g(t, U_t)$ depending on the current state of the system (t, U_t) and a random variable z independent of the current state following the relationship

$$\xi_i = g(\tau_i, U_{\tau_i})\, z, \qquad i = 0, 1, \ldots. \tag{3.10}$$

In order to write this in a more compact form we assume that $\{q_t, t \geq s\}$ is a Poisson process independent of the d-dimensional Brownian motion $\{B_t, t \geq s\}$. It is defined as

$$
dq_t = \begin{cases} z & \text{with probability } \lambda_t dt, \\ 0 & \text{with probability } 1 - \lambda_t dt, \end{cases}
\tag{3.11}
$$

where λ_t is the *mean arrival rate* of a Poisson jump.

Let the amplitude of the jump $z \in \mathbb{R}^n$ be random with density function $\varphi(z)$, i.e., $\varphi(z)dz$ is the probability of a jump amplitude vector contained in the Borel subset $(z, z + dz)$ of $\mathbb{R}^n$. Therefore, we state equation (3.9) as

$$
\begin{aligned}
dU_t &= \mu(t, U_t)\, dt + \sigma(t, U_t)\, dB_t + g(t, U_t)\, dq_t, \qquad t \geq s, \\
U_s &= u,
\end{aligned}
\tag{3.12}
$$

---

[30]We set $\tau_0 = s$.

where $g(t, U_t) \in \mathbb{R}^n$ is a known nonrandom function satisfying again certain linear growth and Lipschitz conditions.

It is sometimes convenient to express (3.12) in terms of an integral representation as:

$$U_t = u + \int_s^t \mu(l, U_l) \, dl + \int_s^t \sigma(l, U_l) \, dB_l + \int_s^t g(l, U_l) \, dq_l. \qquad (3.13)$$

After having defined the jump-diffusion process and before formulating the corresponding impulse control problem we will demonstrate how to compute a functional of the above defined process U_t, which will be necessary to solve stochastic control problems in a real option setting.

Computing a Functional: In order to solve stochastic control problems it is often necessary to compute functions that depend on a certain number of deterministic and stochastic variables. Functions of the following form are especially important in economic analyses:

$$\Phi(s, u) = \mathbb{E}\left[\int_s^T f(t, U_t) \, dt\right] \qquad (3.14)$$

where U_t follows a mixed jump-diffusion process as in (3.12) and $\mathbb{E}$ denotes expectation with respect to the probability measure $\mathbb{P}$. The function $f(t, U_t)$ can be interpreted as a profit or cash flow rate that a firm earns continuously. (3.14) can be expanded using a generalized version of Itô's lemma[31] which will be explained shortly. If we consider the infinitesimal time interval $(s, s + dt)$, $\Phi(s, u)$ can be written as

$$\Phi(s, u) = \mathbb{E}\left[\int_s^{s+dt} f(t, U_t) \, dt\right] + \mathbb{E}\left[\int_{s+dt}^T f(t, U_t) \, dt\right] \qquad (3.15)$$

where the first summand on the right side simplifies to

$$\mathbb{E}\left[\int_s^{s+dt} f(t, U_t) \, dt\right] = f(s, u) dt \qquad (3.16)$$

---

[31] See Harrison [87], p. 54ff, for an intuitive explanation of Itô's lemma or Øksendal [180], p. 32ff, for a more rigorous mathematical treatment of the subject.

as $dt \to 0$. The second summand is more complicated because one has to distinguish whether a jump has occurred in the time interval $(s, s + dt)$ or not.

$$
\mathbb{E}\left[\int_{s+dt}^{T} f(t, U_t)\, dt\right] \tag{3.17}
$$

$$
= \mathbb{E}^q\left[\underbrace{\mathbb{E}\left[\int_{s+dt}^{T} f(t, U_t)\, dt\,\middle|\, \text{jump occurs in } (s, s + dt)\right]}_{(I)}\right.
$$

$$
\left. + \underbrace{\mathbb{E}\left[\int_{s+dt}^{T} f(t, U_t)\, dt\,\middle|\, \text{no jump occurs in } (s, s + dt)\right]}_{(II)}\right]
$$

where $\mathbb{E}^q$ denotes expectation with respect to the Poisson process q_t. As defined above the probability of a jump in an infinitesimal interval $(s, s + dt)$ is $\lambda_t dt$ with jump size density function $\varphi(z)$. If a jump occurs in this interval the process U_t jumps from u to $u + g(s, u)z$. Since z is a random variable we have to average over all possible outcomes of the jump size z. Consequently, (I) becomes

$$
(I) \;=\; \int_{\mathbb{R}^n} \Phi(s, u + g(s, u)z)\, \varphi(z)\, dz. \tag{3.18}
$$

In case of no jump the underlying stochastic process behaves like an Itô diffusion and Itô's lemma can be applied to get

$$
(II) \;=\; \Phi(s, u) \tag{3.19}
$$

$$
+ \left[\frac{\partial \Phi}{\partial s} + \sum_{i=1}^{n} \mu_i \frac{\partial \Phi}{\partial u_i} + \frac{1}{2} \sum_{i,j=1}^{n} (\sigma \sigma^T)_{i,j} \frac{\partial^2 \Phi}{\partial u_i \partial u_j}\right] dt
$$

with σ^T the transpose of matrix σ and $(\cdot)_{ij}$ $(i, j = 1, \ldots, n)$ the (i, j)th element of the $n \times d$-matrix $\sigma \sigma^T$. Taking now expectation with respect to q_t in (3.17) yields

$$
\mathbb{E}\left[\int_{s+dt}^{T} f(t, U_t)\, dt\right] = \lambda_s dt\, (I) + (1 - \lambda_s dt)\, (II) \tag{3.20}
$$

$$
= \lambda_s dt \int_{\mathbb{R}^n} \Phi(s, u + g(s, u)z)\, \varphi(z)\, dz + (1 - \lambda_s dt)
$$

$$
\times \left[\Phi(s, u) + \left[\frac{\partial \Phi}{\partial s} + \sum_{i=1}^{n} \mu_i \frac{\partial \Phi}{\partial u_i} + \frac{1}{2} \sum_{i,j=1}^{n} (\sigma \sigma^T)_{i,j} \frac{\partial^2 \Phi}{\partial u_i \partial u_j}\right] dt\right].
$$

Substituting (3.20) together with (3.16) into (3.15) and ignoring terms of order higher than dt gives the following equation for $\Phi(s, u)$:

$$\Phi(s, u) = \Phi(s, u) + \left[f(s, u) + \lambda_s \int_{\mathbb{R}^n} \Phi(s, u + g(s, u)z)\, \varphi(z)\, dz \right. \quad (3.21)$$

$$\left. -\lambda_s \Phi(s, u) + \frac{\partial \Phi}{\partial s} + \sum_{i=1}^{n} \mu_i \frac{\partial \Phi}{\partial u_i} + \frac{1}{2} \sum_{i,j=1}^{n} (\sigma\sigma^T)_{i,j} \frac{\partial^2 \Phi}{\partial u_i \partial u_j} \right] dt.$$

For this to be true the expression in brackets must equal zero. Hence $\Phi(s, u)$ needs to be the solution of the integro-partial differential equation (PDE)

$$\underbrace{\frac{\partial \Phi}{\partial s} + \sum_{i=1}^{n} \mu_i \frac{\partial \Phi}{\partial u_i} + \frac{1}{2} \sum_{i,j=1}^{n} (\sigma\sigma^T)_{i,j} \frac{\partial^2 \Phi}{\partial u_i \partial u_j} + f(s, u)}_{\mathcal{A}\Phi(s,u)} \quad (3.22)$$

$$+\lambda_s \int_{\mathbb{R}^n} [\Phi(s, u + g(s, u)z) - \Phi(s, u)]\, \varphi(z)\, dz = 0.$$

This is the generalized Itô formula. $\mathcal{A}$ is often called the *operator* of the Itô diffusion without jump component. The derivation of a solution to this equation together with the corresponding boundary condition (e.g., $\Phi(T, u) = 0$ in the simplest case) usually requires numerical techniques.

Note, that in case of a deterministic jump size z the integro PDE reduces to the PDE

$$\mathcal{A}\Phi(s, u) + f(s, u) + \lambda_s [\Phi(s, u + g(s, u)z) - \Phi(s, u)] = 0. \quad (3.23)$$

Furthermore, if we consider the discounted problem

$$\Phi(s, u) = \mathbb{E}\left[\int_s^T e^{-r(t-s)} f(t, U_t)\, dt \right] \quad (3.24)$$

where r denotes a certain discount rate, the integro-PDE (3.22) only slightly changes to

$$\mathcal{A}\Phi(s, u) + f(s, u) - r\Phi(s, u) \quad (3.25)$$

$$+ \lambda_s \int_{\mathbb{R}^n} [\Phi(s, u + g(s, u)z) - \Phi(s, u)]\, \varphi(z)\, dz = 0.$$

3.3 Impulse Control Model for Valuing Real Options

3.3.1 Problem Formulation

After the definition of the underlying state process that contains the source of uncertainty the decision maker is exposed to, we will give a formal exposition of the way this state process can be influenced via impulse control techniques. In fact, one can distinguish two cases. The first case covers problems where the exercise of an impulse control has a *direct impact* on the stochastic process. This is, for example, the case when the decision maker has means to directly alter or shift the underlying stochastic variables in a certain sense. An example in the literature is the control of exchange rates by federal banks that intervene in the currency markets from time to time by buying or selling large amounts in order to force exchange rates to stay in a certain band.[32]

In the second case, we assume the underlying stochastic processes to be exogenously given to the decision maker, who is not able to change the stochastic process itself but has the opportunity to adjust certain decision variables to the stochastically changing environment. We will call this case the *indirect impact* case. In order to specify the system to be controlled the state space will then be determined by the current state of the stochastic processes and in addition the current values of the decision variables the decision maker can control.

While the stochastic nature of most real option applications is exogenously given, the indirect impact case will mainly be the modelling setting which is employed for the problems considered here, but we will present the main results also taking direct controls into account. A classical example may be the valuation of a flexible manufacturing system where the price and cost processes are exogenously given and management has yet to decide about its operating policy. In this setting the operating policy may consist of the choice of products to manufacture and the corresponding quantities under capacity restrictions. The expanded state space is then determined by the exogenous stochastic price and cost processes and the current production rates. Since adjustments in the operating policy raise switching costs which comprise fixed and variable parts, an impulse control strategy is obviously the right choice.

We will now start to develop a very general impulse control model which serves as a building block for the following valuation of interacting real options. Since it represents a main part of the approach taken in this study, the model is formally developed and proofs of the main theorems are given. The basic idea of impulse control has already been given. After the presentation of the generic impulse control model, an economically intuitive framework for pricing complex real option interactions will be established.

To start with, suppose a value maximizing decision maker has to make an — at least partially — irreversible investment decision in an uncertain environment. The decision maker has the means to manage the investment by several decision variables. Each time he changes the decision variables and therefore controls the system directly or indirectly, costs are incurred that contain a fixed positive amount.

Consider a value maximizing firm that has the opportunity to control several variables in its projects. The set of decision variables the firm can alter is called *action space* $\mathcal{Z}$. The action space $\mathcal{Z}$ is further divided into the set of decision

---

[32]See Jeanblanc-Pique [105].

variables that have a direct impact on the firm's stochastic environment denoted by $\mathcal{Z}^d$ and its complement $\mathcal{Z}^i$ which contains all decision variables which can only be indirectly adjusted to the changing environment. Furthermore let Z_t denote the vector of the state of the decision variables at time t. Thus we have for all t,

$$Z_t \in \{z_1, \ldots, z_k\} =: \mathcal{Z} = \mathcal{Z}_d \cup \mathcal{Z}^i \subseteq \mathbb{R}^k \qquad (3.26)$$

where Z_t is assumed to be right continuous with left limits. Furthermore we model the firm's environment U_t at time t as a stochastic process in $\mathbb{R}^n$ satisfying the stochastic differential equation (SDE)

$$dU_t = b(t, U_t, Z_t)\, dt + \sigma(t, U_t, Z_t)\, dB_t \qquad (3.27)$$

where B_t denotes d-dimensional Brownian motion. The functions $b : \mathbb{R}^{n+1} \times \mathcal{Z} \to \mathbb{R}^n$ and $\sigma : \mathbb{R}^{n+1} \times \mathcal{Z} \to \mathbb{R}^{n \times d}$ satisfy the following assumptions:[33]

- b and σ are twice continuously differentiable,

- b and σ are measurable functions satisfying for all $U_t \in \mathbb{R}^n$ and $t \in [0, \infty)$,

$$|b(t, U_t, Z_t)| + |\sigma(t, U_t, Z_t)| \leq C(1 + |U_t|) \qquad (3.28)$$

with $Z_t \in \mathcal{Z}$ for some constant C, where $|\sigma|^2 = \sum |\sigma_{ij}|^2$ and such that

$$|b(t, U, Z_t) - b(t, U', Z_t)| + |\sigma(t, U, Z_t) - \sigma(t, U', Z_t)| \qquad (3.29)$$
$$\leq D|U - U'|;$$

with $U, U' \in \mathbb{R}^n, t \in [0, \infty)$ and $Z_t \in \mathcal{Z}$ for some constant D.

These are the so-called *Lipschitz and growth conditions*. Then the state of the whole economic system at time t consists of the realization of the environmental uncertainty and the value of the decision variables of the firm. It is given by the stochastic process

$$X_t = \begin{pmatrix} t \\ U_t \\ Z_t \end{pmatrix}. \qquad (3.30)$$

The probability law of X_t given an initial value $X_0 = x = (s, u, z)$ is denoted by $\mathbb{P}^x$ and expectation with respect to $\mathbb{P}^x$ is denoted by $\mathbb{E}^x$.

An *impulse control strategy* for this system is defined[34] as the sequence

$$w = \{(\theta_k, \zeta_k); k \in \mathbb{N}\} \qquad (3.31)$$

---

[33] See Øksendal [180], p. 48f.
[34] See Korn [121].

of intervention times θ_k and actions (or impulses) $\zeta_k = Z_{\theta_k}$ with

(i) $s \leq \theta_k \leq \theta_{k+1}$ a.s. $\forall k \in \mathbb{N}$,

(ii) θ_k is a stopping time w.r.t. the filtration $\{\mathcal{F}_t\}$ generated by the Brownian motion $\{B_t\}$i.e.,

$$\sigma\left((X_{s-}), s \leq t;\ (\theta_n, \zeta_n), n < k\right),$$

(iii) ζ_k are measurable w.r.t. $\sigma((X_{\theta_k-}), (\zeta_n), n < k)$,   (3.32)

(iv) $\mathbb{P}(\lim_{k\to\infty} \theta_k \leq T) = 0;$ $\forall T \geq 0,$

(v) $\zeta_k \in \mathcal{Z}.$

We can regard $\theta_1, \theta_2, \ldots$ as the times when it is decided to interfere in the system and the corresponding new decision variables of the firm are chosen to be $\zeta_1, \zeta_2, \ldots$. Note that (iv) of the definition of an impulse control strategy assures that the set of impulse times is countable. If an impulse control satisfies (iv) we call it *admissible*.

Let $\mathcal{W}$ be the set of all admissible impulse control strategies w. If we apply an impulse control strategy $w \in \mathcal{W}$ to the system, it takes the form

$$X_t = X_t^{(w)} = \begin{pmatrix} t \\ U_t^{(w)} \\ \zeta_k \end{pmatrix} \qquad \text{if } \theta_k \leq t \leq \theta_{k+1}. \qquad (3.33)$$

While an impulse control may also exist as a part which directly alters the underlying stochastic process U_t, it is a function of the impulse control w applied to the system. Therefore, we have to write $U_t = U_t^{(w)}$. To specify this further assume that $\zeta_k^d \in \mathcal{Z}^d$ is the part of the impulse which directly influences U_t. The impact of ζ_k^d on U_t can be described by a given measurable function

$$\gamma : \mathbb{R}^{n+1} \times \mathcal{Z} \times \mathcal{Z}^d \to \mathbb{R}^n. \qquad (3.34)$$

Thus, $U_t^{(w)}$ has the integral representation

$$U_t^{(w)}(\omega) = u + \int_s^t b(l, U_l^{(w)}, Z_l)\, dl + \int_s^t \sigma(l, U_l^{(w)}, Z_l)\, dB_l(\omega)$$
$$+ \sum_{k:\theta_k \leq t} \gamma(X_{\theta_k^-}^{(w)}, \zeta_k) \qquad (3.35)$$

where

$$X_{\theta_k^-}^{(w)} = \lim_{t\uparrow\theta_k} X_t = (\theta_k, \lim_{t\uparrow\theta_k} U_t^{(w)}, \zeta_{k-1}). \qquad (3.36)$$

Note, that $X_t^{(w)}$ is right continuous for all $w \in \mathcal{W}$. We will drop the superscript (w) from now on whenever it is clear from the context.

Running the system yields an instantaneous cash flow $f(x)$ when the system is in state x. For $x = (s, u, z) \in \mathbb{R}^{n+1} \times \mathcal{Z}$ and $\zeta \in \mathcal{Z}$ let $H(x, \zeta) \in \mathbb{R}$ be the cost of switching the system from z to ζ when the state of the system is $x = (s, u, z)$. Furthermore, assume that there is no impulse control $w \in \mathcal{W}$ which generates infinite profits, i.e.,

$$\mathbb{E}^x \left[\int_s^\infty \left| f(X_t^{(w)}) \right| dt \right] \leq \infty \qquad \forall x, \forall w \in \mathcal{W}. \tag{3.37}$$

Then we calculate the expected value of the firm when the impulse control strategy $w = (\theta_1, \theta_2, \ldots; \zeta_1, \zeta_2 \ldots) \in \mathcal{W}$ is applied to the system by

$$J^w(x) = \mathbb{E}^x \left[\int_s^\infty f(X_t) \, dt - \sum_{j=1}^\infty H(X_{\theta_j^-}, \zeta_j) \mathbb{I}_{\{\theta_j < \infty\}} \right]. \tag{3.38}$$

The switching cost function $H : \mathbb{R}^{n+1} \times \mathcal{Z} \times \mathcal{Z} \to \mathbb{R}^+$ satisfies

$$
\begin{array}{ll}
(i) & H(x, \zeta) > 0 \qquad \forall x \in \mathbb{R}^{n+1} \times \mathcal{Z} \text{ and all } \zeta \neq z; \\
(ii) & (s, u) \to H(s, u, z, \zeta) \quad \text{is continuous for all } z, \zeta; \\
(iii) & H(s, u, z, \zeta_2) \leq H(s, u, z, \zeta_1) + H(s, u^{\zeta_1}, \zeta_1, \zeta_2) \\
& \text{if } z \neq \zeta_1 \neq \zeta_2 \neq z.
\end{array}
\tag{3.39}
$$

(i) means that we have a positive fixed cost component when switching to another state. In order for Z_t to be right continuous, we have to impose (iii). It says that a direct switch from z to ζ_2 cannot be more expensive than first switching to an intermediate ζ_1 and from there to ζ_2. It is easy to see that this condition is just the triangle inequality.[35] If (iii) was not satisfied, we would have two instantaneous switches at the same time which establishes that Z_t could not be right continuous.

We are now ready to state the impulse control problem as follows:

General Impulse Control Problem
Find for all $x = (s, u, z)$,

$$\tilde{\phi}(x) := \sup_{w \in \mathcal{W}} J^w(x) \tag{3.40}$$

and an optimal impulse control $\tilde{w}$, i.e., $\tilde{w} \in \mathcal{W}$ such that

$$\tilde{\phi}(x) = J^{\tilde{w}}(x). \tag{3.41}$$

---

[35]Note, that if the inequality in (iii) is strict, then two subsequent impulse control times θ_k and θ_{k+1} can never coincide, since first switching to an intermediate state ζ_1 before switching to ζ_2 can never be optimal. Therefore we have $\theta_k < \theta_{k+1}$ in this case.

Unfortunately, it is not possible to directly compute the function $\tilde{\phi}$, at least not for problems formulated in such a general way as it is done here. The usual procedure with this type of stochastic control problems is to find a solution candidate for a given set of decision variables $\mathcal{Z}$, specified stochastic processes U_t, and specific given functions γ, f and H. Of course, the way to find a solution for a specific problem formulation depends crucially on the choice of the input processes and functions. Solutions may be obtained in closed form or only by employing numerical solution techniques. We will give examples for both in the course of the chapter. The next step is then to prove that the found solution candidate is in fact the unique optimal solution of the impulse control problem. This is done here for the very general impulse control problem through a so-called *verification theorem*. The verification theorem simply states that if a solution candidate $\hat{\phi}$ fulfills the conditions stated in the verification theorem, then it is the unique optimal solution of the impulse control problem. Thus, the conditions of the verification theorem are sufficient for the solution of the general impulse control problem.

3.3.2   Impulse Control Verification Theorem

In order to check the optimality of a solution candidate $\hat{\phi}$ we have to determine a system of *sufficient quasi-variational inequalities*[36] which any solution ϕ has to satisfy. Using some advanced techniques of dynamic programming, we will state conditions for time intervals where the system evolves freely and conditions for time instances where intervention takes place and an impulse control is applied. Afterwards we are able to state a verification theorem which relates the solution of the general impulse control problem to the solution of a system of quasi-variational inequalities. This important result will then be proved using the concept of *stochastically C^2-functions* and a generalized Dynkin formula introduced by Brekke/Øksendal [32].

We use the following argument. When the system is in state X_t at time t, we can either wait a certain period of time before applying an impulse control, or intervene immediately.

In the first case, let Y_t represent the state of the system during the time interval where no intervention takes place. So for fixed $z \in \mathcal{Z}$ we have the $(n + 1)$-dimensional process

$$Y_t = (t, U_t, z) = (t, U_t), \qquad t \in [\theta_k, \theta_{k+1}). \tag{3.42}$$

Then Y_t is an Itô diffusion whose generator is given by[37]

$$\mathcal{A} \; = \; \frac{\partial}{\partial s} + \sum_{i=1}^{n} b_i \frac{\partial}{\partial u_i} + \frac{1}{2} \sum_{i,j=1}^{n} (\sigma \sigma^T)_{i,j} \frac{\partial^2}{\partial u_i \partial u_j} \tag{3.43}$$

---

[36]See Bensoussan/Lions [20] for a theoretic introduction into impulse control problems and the role of quasi-variational inequalities for their solution.

[37]This follows directly from applying Itô's lemma to an arbitrary measurable function $g(t, U_t)$ that depends on t and U_t.

with σ^T the transpose of matrix σ and $(\cdot)_{ij}$ $(i, j = 1, \ldots, n)$ the (i, j)th element of the $n \times n$-matrix $\sigma \sigma^T$.

We see that all candidates for the value function ϕ must at least be differentiable with respect to t and twice differentiable with respect to u in the region where it is not optimal to intervene in the system. Let this region which is usually called *continuation region* be denoted by C.[38]

On the other hand, when X_t is in the *intervention region* $\bar{C}$, which is the complement of the continuation region, then an impulse control is applied right away and the optimal action is chosen. Therefore, we define the following maximum operator $\mathcal{M}$ as

$$\mathcal{M}\phi := \max_{\zeta \in \mathcal{Z} \backslash \{z\}} \left\{ \phi(s, u + \gamma(x, \zeta^d), \zeta) - H(x, \zeta) \right\} \tag{3.44}$$

for all Borel functions $\phi : \mathbb{R}^{n+1} \times \mathcal{Z} \to \mathbb{R}$. This means that whenever the process X_t reaches the boundary $\partial C \subset \bar{C}$ of the continuation region the optimal switch according to the maximum operator is applied to X_t, which brings X_t back into the continuation region where it evolves freely.

From the definition of the maximum operator $\mathcal{M}$ it is clear that the value function is at least continuous and differentiable with respect to t and u at the boundary of the continuation region ∂C. But twice differentiability with respect to u which is necessary to apply Itô's formula and Bellman's optimality principle, may not be satisfied. In order to assure the smoothness of the value function at the boundary between interference and continuation region ∂C, we make use of the following definition:[39]

Definition
A continuous function $\varphi : \mathbb{R}^{n+1} \to \mathbb{R}$ is called *stochastically* C^2 in a domain $D \subset \mathbb{R}^{n+1}$ with respect to the Itô diffusion Y_t if the following generalized Dynkin formula holds:

$$\mathbb{E}^y \left[\varphi(Y_{\theta'}) | \mathcal{F}_\theta \right] = \varphi(Y_\theta) + \mathbb{E}^y \left[\int_\theta^{\theta'} \mathcal{A}\varphi(Y_t) \, dt | \mathcal{F}_\theta \right] \tag{3.45}$$

for all stopping times $\theta \leq \theta' \leq \inf\{t > 0; Y_t \notin D\} < \infty$.
Remark:

(i) In Brekke/Øksendal [32] it was proved that a function φ which is continuously differentiable, i.e., C^1, everywhere and twice continuously differentiable, i.e., C^2, outside a "thin" set (with respect to the Green measure

---

[38]The concrete form of the continuation region C will be specified later.
[39]See Brekke/Øksendal [32], p. 192ff, and Mundaca/Øksendal [175], p. 232.

$G^x(\cdot)$ of X_t) is stochastically C^2. This corresponds to the "value matching" and "smooth pasting" condition used for example in Dixit/Pindyck [72] to determine optimal entry and exit barriers of a value maximizing firm.

(ii) The generalized Dynkin formula will be essential for the proof of the following theorem which will state the main result of this section.

Impulse Control Verification Theorem

(1) Suppose there exists a continuous function $\phi(s, u, z)$ in $\mathbb{R}^{n+1}$ satisfying the following properties:

$$(i) \qquad \phi \text{ is stochastically } C^2 \text{ w.r.t. } Y_t \text{ in } C, \qquad\qquad (3.46)$$

$$(ii) \qquad \{\phi(\tau, U_\tau, z)\}_{\tau \in \mathcal{T}} \text{ is uniformly integrable w.r.t. } \mathbb{P}^y \qquad (3.47)$$
$$\text{for all } z \in \mathcal{Z} \text{ and } \tau \in \mathcal{T} \text{ where } \mathcal{T} \text{ is the set of all}$$
$$\mathcal{F}_t\text{-stopping times,}$$

$$(iii) \qquad \phi \geq \mathcal{M}\phi \text{ everywhere,} \qquad\qquad\qquad\qquad (3.48)$$

$$(iv) \qquad \mathcal{A}\phi + f \leq 0 \text{ almost everywhere w.r.t. the Green} \qquad (3.49)$$
$$\text{measure } G^y(\cdot).$$

Then

$$\phi(x) \quad \geq \quad J^w(x) \qquad \forall w \in \mathcal{W} \quad \text{and all } x = (s, u, z). \qquad (3.50)$$

(2) Suppose that in addition

$$\mathcal{A}\hat{\phi} + f = 0 \qquad\qquad\qquad (3.51)$$

holds for all $x = (s, u, z) \in C$ where

$$C := \left\{ x = (s, u, z);\ \hat{\phi}(s, u, z) > \mathcal{M}\hat{\phi}(s, u, z) \right\}. \qquad (3.52)$$

Define the following impulse control strategy $\hat{w} = (\hat{\theta}_1, \hat{\theta}_2, \ldots; \hat{\zeta}_1, \hat{\zeta}_2, \ldots)$ inductively as follows: Put $\hat{\theta}_0 = 0$ and

$$\hat{\theta}_{k+1} = \inf \left\{ t > \hat{\theta}_k;\ X_t^{(k)} \notin C \right\}, \qquad k = 0, 1, 2, \ldots \qquad (3.53)$$

where $X_t^{(k)}$ is the result of applying the control

$$(\hat{\theta}_1, \hat{\theta}_2, \ldots, \hat{\theta}_k, \infty; \hat{\zeta}_1, \hat{\zeta}_2, \ldots, \hat{\zeta}_k, \cdots) \qquad\qquad (3.54)$$

to X_t and choose $\hat{\zeta}_{k+1}$ such that

$$\mathcal{M}\hat{\phi}\left(\hat{\theta}_{k+1}, U_{\hat{\theta}_{k+1}}, \hat{\zeta}_{k+1}\right) = \qquad\qquad (3.55)$$
$$\hat{\phi}\left(\hat{\theta}_{k+1}, U_{\hat{\theta}_{k+1}^-} + \gamma(X_{\hat{\theta}_{k+1}^-}, \hat{\zeta}_{k+1}^d), \hat{\zeta}_{k+1}\right) - H\left(X_{\hat{\theta}_{k+1}^-}, \hat{\zeta}_{k+1}\right).$$

Then

$$\hat{\phi} = J^{\hat{w}}(x) = \tilde{\phi}(x) \tag{3.56}$$

and the optimal impulse control strategy is given by

$$\hat{w} = \tilde{w}. \tag{3.57}$$

Proof of(1):[40]
Using the generalized Dynkin formula applied to $\phi(s, u, z)$ where $y = (s, u, z)$ for fixed $z \in \mathcal{Z}$ and letting $s \le \theta \le \theta' \le \infty$ be two stopping times, we have

$$\mathbb{E}^y \left[\phi(\theta', U_{\theta'}, z) | \mathcal{F}_\theta \right] = \tag{3.58}$$

$$\phi(\theta, U_\theta, z) + \mathbb{E}^y \left[\int_\theta^{\theta'} \mathcal{A}\phi(t, U_t, z) \, dt | \mathcal{F}_\theta \right]$$

and by (3.49) ($\mathcal{A}\phi \le -f$) it follows that

$$\phi(\theta, U_\theta, z) \ge \mathbb{E}^y \left[\int_\theta^{\theta'} f(Y_t) \, dt + \phi(\theta', U_{\theta'}, z) | \mathcal{F}_\theta \right]. \tag{3.59}$$

Next let $w = (\theta_1, \theta_2, \ldots)$ be an impulse control with $\theta_1 > s$ and let $X_t = X_t^{(w)} = (t, U_t, Z_t)$ be the w-controlled process. By (3.48) ($\phi \ge \mathcal{M}\phi$) the following inequality holds for $\theta = \theta_k, \theta' = \theta_{k+1}$ and $z = Z_{\theta_k}; k = 0, 1, 2, \ldots$:

$$\phi(\theta_k, U_{\theta_k}, Z_{\theta_k}) \ge \mathbb{E}^y \left[\int_{\theta_k}^{\theta_{k+1}} f(X_t) \, dt + \phi(\theta_{k+1}, U_{\theta_{k+1}^-}, Z_{\theta_k}) | \mathcal{F}_{\theta_k} \right]$$

$$\ge \mathbb{E}^y \left[\int_{\theta_k}^{\theta_{k+1}} f(X_t) \, dt \right. \tag{3.60}$$

$$+ \phi \left(\theta_{k+1}, U_{\theta_{k+1}^-} + \gamma(X_{\theta_{k+1}^-}, \zeta_{k+1}), \zeta_{k+1} \right)$$

$$\left. - H \left(X_{\theta_{k+1}^-}, \zeta_{k+1} \right) | \mathcal{F}_{\theta_k} \right]$$

where we used the fact that

$$X_{\theta_{k+1}^-} = \left(\theta_{k+1}, U_{\theta_{k+1}^-}, Z_{\theta_k} \right) \tag{3.61}$$

---

[40] See especially Bensoussan/Lions [20], p. 615–634, and Øksendal [180], chapter X.

and

$$\phi\left(X_{\theta_{k+1}^-}\right) \geq \mathcal{M}\phi\left(\theta_{k+1}, U_{\theta_{k+1}^-}, Z_{\theta_k}\right) \tag{3.62}$$

$$= \phi\left(\theta_{k+1}, U_{\theta_{k+1}^-} + \gamma(X_{\theta_{k+1}^-}, \zeta_{k+1}), \zeta_{k+1}\right)$$

$$- H\left(X_{\theta_{k+1}^-}, \zeta_{k+1}\right)$$

and

$$Z_{\theta_{k+1}} = \zeta_{k+1} \qquad \text{by (3.33).} \tag{3.63}$$

(3.60) essentially means that the value function ϕ, calculated at the last stopping time θ_k if w is applied to the system, is at least as large as the reward earned by the uncontrolled system up to the next intervention time θ_{k+1} plus the value of the system when the optimal impulse control is applied at time θ_{k+1} minus switching costs for changing the state of the system. Since we are interested in the value of the system at starting point $y = (s, u, z)$ up to infinity, we sum up over all stopping times from $k = 0$ to $k = n - 1$, take expectation with respect to y and let n go to infinity. Therefore, we have

$$\phi(s, u, z) + \sum_{k=1}^{n-1} \mathbb{E}^y\left[\phi(\theta_k, U_{\theta_k}, Z_{\theta_k})\right] \tag{3.64}$$

$$\geq \mathbb{E}^y\left[\int_s^{\theta_n} f(X_t)\, dt - \sum_{k=1}^n H(X_{\theta_k^-}, Z_{\theta_k}) + \sum_{k=1}^n \phi(\theta_k, U_{\theta_k}, Z_{\theta_k})\right]$$

and consequently

$$\phi(s, u, z) \tag{3.65}$$

$$\geq \mathbb{E}^y\left[\int_s^{\theta_n} f(X_t)\, dt - \sum_{k=1}^n H(X_{\theta_k^-}, Z_{\theta_k}) + \phi(\theta_n, U_{\theta_n}, Z_{\theta_n})\right].$$

With $n \to \infty$, also $\theta_n \to \infty$ and by assuming

$$\phi(s, U_t, z) \xrightarrow{t \to \infty} 0 \qquad \text{almost surely } \mathbb{P}^y \quad \forall z \in \mathcal{Z}, y = (s, u, z), \tag{3.66}$$

which is usually the case for problems considered in real option pricing due to appropriate discounting, we get

$$\phi(s, u, z) \geq \mathbb{E}^y\left[\int_s^\infty f(X_t)\, dt - \sum_{k=1}^\infty H(X_{\theta_k^-}, Z_{\theta_k})\right] = J^w(y) \tag{3.67}$$

and the proof of (1) is completed.

Proof of (2):
Applying the optimal intervention times $\hat{\theta}_k$ and $\hat{\theta}_{k+1}$ to equation (3.65) and noting that we have equality $\mathcal{A}\phi = -f$ between optimal intervention times, we get equality in (3.65). Thus, we get

$$\hat{\phi}(s, u, z) \tag{3.68}$$
$$= \mathbb{E}^y \left[\int_s^{\hat{\theta}_n} f(X_t)\, dt - \sum_{k=1}^{n} H(X_{\hat{\theta}_k^-}, Z_{\hat{\theta}_k}) + \hat{\phi}(\hat{\theta}_n, U_{\hat{\theta}_n}, Z_{\hat{\theta}_n}) \right].$$

Furthermore, by letting n go to infinity and remembering that by definition of an admissible impulse control (3.32-iv) holds and also (3.37), we see that

$$\hat{\theta}_n \stackrel{n \to \infty}{\longrightarrow} \infty \tag{3.69}$$

and $\hat{w} \in \mathcal{W}$. As a consequence the summed switching costs do not explode and the right side of equation (3.68) is finite.

This shows in fact that

$$\hat{\phi}(x) \quad = \quad J^{\hat{w}}(x) \tag{3.70}$$

and while

$$\hat{\phi}(x) \quad \geq \quad J^w(x) \tag{3.71}$$

it follows that indeed

$$\hat{\phi}(x) \quad = \quad \sup_{w \in \mathcal{W}} J^w(x) = \tilde{\phi}(x) \tag{3.72}$$

is the optimal value function with optimal impulse control strategy $\tilde{w} = \hat{w}$.

Remark:

(i) To state the main result of the theorem in a more compact form we have just proved that (similar to Section 3.2.2) the optimal value function ϕ can be obtained as the solution to the following set of sufficient quasi-variational inequalities (qvi):

(1) $\mathcal{A}\phi(x) + f(x) \leq 0,$

(2) $\phi(x) \geq \mathcal{M}\phi(x),$ $\tag{3.73}$

(3) $(\phi(x) - \mathcal{M}\phi(x))\, (\mathcal{A}\phi(x) + f(x)) = 0.$

If a solution to these inequalities can be found, it is also the optimal solution to the original impulse control problem (3.40)–(3.41). Furthermore, it can be shown that (3.73) defines ϕ in a unique way.[41]

(ii) Although the above theorem does not yield the solution of the impulse control problem directly, it can help us in two ways. First, if a solution candidate is explicitly given it is an easy task to check whether it fulfills the required conditions of the theorem. But still some intuition is needed to derive closed form solutions. If on the other hand a numerical solution technique is employed, we can use the set of qvi in (3.73) to construct a numerical solution such that the conditions of the impulse control verification theorem hold.

3.3.3 Interpretation and Extensions

We have presented a very general model of impulse control where two different types of interventions were qualitatively distinguished. Namely, interventions which directly alter the underlying stochastic process noted by the set $\mathcal{Z}^d$ and its complement $\mathcal{Z}^i = \mathcal{Z} \setminus \mathcal{Z}^d$, which contains the part of the action space that only influences the payoff functions in reaction to changes of the stochastic processes but not the underlying stochastic processes themselves. One can think of the following example in terms of real option pricing. Consider a company that produces a single product whose demand is stochastic and follows an Itô diffusion. The firm may have only two sources of flexibility to react on changes in the aggregate demand for its product. The first is to react without changing the stochastic demand by making its entry and exit decisions contingent on the realization of the current demand. Clearly, the impulse control to enter or exit the market is indirect and belongs to $\mathcal{Z}^i$ since the decision does not change the demand for the product itself but the state of the whole system. If on the other hand the firm has also the flexibility to change the demand itself, as is the case for example with a marketing campaign that alters the demand level in a certain way, the demand process is directly and endogenously changed by the firm by exerting a direct control. While the indirect kind of flexibility is the one usually considered in real option pricing, the direct control case is at least as relevant from a managerial point of view. The problem with the latter case is the endogeneity of the stochastic underlying which makes contingent claims analysis quite difficult due to market incompleteness.

Albeit a very general model, several extensions can be thought of. Changes of the general impulse control model concerning the following points could be handled:

- state dependent action space $\mathcal{Z}$,

- finite time horizon T,

---

[41] See Bensoussan/Lions [20] and Korn [120].

- discounting with discount rate ρ, and

- stochastic processes U_t including Poisson components.

First of all, note that the proof of the theorem is independent of the form of the action space $\mathcal{Z}$. Hence, it is also valid if $\mathcal{Z}$ is a function $\mathcal{Z}(u, z)$ of the controlled process. This statement holds because the impulse controls need to be reversible. It does not hold if $\mathcal{Z}$ is a direct function of time t since reversibility is not satisfied. This case will be subject to the generalized timing option in (3.4).

Nothing would dramatically change if we considered an impulse control model with finite planning horizon T instead of the perpetual system. The same proof would go through with only some minor changes. The only information needed to be added to the qvi system (3.73) is a terminal condition at time T which may be either a constant or the value function of a subsequent optimization problem. Hence, we can state the terminal condition as

$$\phi^a(T, U_T^a, Z_T^a) = \phi^p(T, U_T^p, Z_T^p) \qquad (3.74)$$

where the superscript a denotes the *ante* optimization problem and p the *post* optimization problem, respectively.

The same holds for the discounted problem. The only change when using a discount rate ρ is in the generator $\mathcal{A}$. We will deal with the discounted problem later in the chapter.

Stochastic processes U_t which include Poisson components in addition to Itô diffusions can as well be integrated into the generalized impulse control model. However, the model had to be modified in two ways. First, which is certainly the easier task, the generator of the value function ϕ had to be changed according to (3.2.2) where we have already computed a functional including Poisson shots. Second, we would have to distinguish cases where the interplay of Poisson jumps and impulse controls yield movements of the controlled state process in the continuation region $\mathcal{C}$ and also in the intervention region $\bar{\mathcal{C}}$. This may complicate the situation from a mathematical point of view without changing the overall results in general. For that reason we will not present mathematical proofs for this case.

To conclude, a first important step towards a stochastic control framework for real options has been established by the generalized impulse control model. Because solution techniques for this kind of models are very case sensitive, they were not presented so far. Instead a verification theorem was formulated and proved that helps to immediately identify the optimality of a solution candidate.

While the generalized impulse control model only captures reversible switches, we will present results for the generalized stopping model which takes care of irreversible switches in the next section, following essentially the same line of arguments as in the above section. Because most classical real options like the option to invest or the option to abandon are assumed to be irreversible, i.e., the opportunity is assumed to be exercisable only once and returning to the previous state is precluded, this is the next step towards an integrated stochastic control model for real options.

3.4   Combined Impulse Control
and Optimal Stopping

In much the same way as defining a generalized impulse control problem we present a generalized model of optimal stopping which will be the second part of the stochastic control framework for real options. Optimal stopping problems are well known in economics and especially in finance, since many American options can be expressed as stopping problems.[42] The basic idea is that a stochastic process may evolve (freely) up to a certain stopping time τ at which an action takes place. In case of an American put, for example, the optimal stopping time τ is the time at which it is optimal to exercise the put. With this in mind it is obvious that an impulse control problem is nothing else than a sequence of optimal stopping problems. The reason that optimal stopping is treated after impulse control models is that we want to add a feature to the model presented here. The following example may illustrate the point. Consider the option to invest in a new plant that will produce a new product. Besides choosing the optimal time to build the plant, management has also the option to choose the capacity of the plant. More abstract, the management has to exercise not only the timing option but also an *intensity option*.[43]

The goal of this section is therefore to develop a generalized optimal stopping model that allows for timing *and* intensity simultaneously.

3.4.1   Problem Formulation

In this part we formulate sufficient variational inequalities for the problem of irreversible switching between two impulse control problems. The model presented here is a generalization of a problem studied by Hu/Øksendal [98]. We assume that stopping takes place between two impulse control problems a and p where, as above, a stands for *ante* and p for *past*, respectively. That is, we formulate in the same way as in the above section two impulse control problems according to two action spaces $\mathcal{Z}^a$ and $\mathcal{Z}^p$ where the processes $Z_t^a \in \mathcal{Z}^a$ and $Z_t^p \in \mathcal{Z}^p$ are right continuous with left limits.

Then the state of the system can be described by the two stochastic processes

$$X_t^a = \begin{bmatrix} t \\ U_t^a \\ Z_t^a \end{bmatrix} \in \mathbb{R}^{n+1} \times \mathcal{Z}^a, \qquad s \le t < \tau \qquad \text{and} \quad (3.75)$$

$$X_t^p = \begin{bmatrix} t \\ U_t^p \\ Z_t^p \end{bmatrix} \in \mathbb{R}^{m+1} \times \mathcal{Z}^p, \qquad \tau \le t \qquad (3.76)$$

---

[42]See Bingham/Kiesel [26], p. 200.

[43]The problem of combined timing and intensity of investment was first considered by Bar-Ilan/Strange [15] for the option to invest.

where the firm's environment at time t is denoted by U_t^a for $t < \tau$ and U_t^p for $t \geq \tau$, respectively. We assume that U_t^a is a stochastic process in $\mathbb{R}^n$ satisfying the following stochastic integral equation for $s \leq t < \tau$:

$$U_t^{(w^a)}(\omega) = u^a + \int_s^t b^a \left(l, U_l^{(w^a)}, Z_l^a \right) dl \qquad (3.77)$$

$$+ \int_s^t \sigma^a \left(l, U_l^{(w^a)}, Z_l^a \right) dB_l(\omega)$$

$$+ \sum_{k:\theta_k \leq t} \gamma^a \left(X_{\theta_k^-}^{(w^a)}, \zeta_k^a \right)$$

for $s \leq t < \tau$ with initial value

$$X_0^a = x^a = (s, u^a, z^a) \qquad (3.78)$$

where $w^a \in \mathcal{W}^a$ is an impulse control policy applied to the stopped impulse control problem at time τ and b^a, σ^a are defined as above. Furthermore, B_t denotes d-dimensional Brownian motion. In much the same way U_t^p is given by

$$U_t^{(w^p)}(\omega) = U_\tau^p + \int_\tau^t b^p \left(l, U_l^{(w^p)}, Z_l^p \right) dl \qquad (3.79)$$

$$+ \int_\tau^t \sigma^p \left(l, U_l^{(w^p)}, Z_l^p \right) dB_l(\omega)$$

$$+ \sum_{k:\theta_k \leq t} \gamma^p \left(X_{\theta_k^-}^{(w^p)}, \zeta_k^p \right)$$

for $\tau \leq t < \infty$ with initial value

$$X_\tau^p = (\tau, U_\tau^p, Z_\tau^p) \qquad (3.80)$$

where $w^p \in \mathcal{W}^p$ is the corresponding impulse control policy starting at time τ in state $(\tau, U_\tau^p, Z_\tau^p)$.

The probability law of X_t^a given $X_0^a = (s, u^a, z^a)$ is denoted by $\mathbb{P}^x$ with expectation $\mathbb{E}^x$.

The impulse control strategies $w^a \in \mathcal{W}^a$ and $w^p \in \mathcal{W}^p$ are defined as in the above section. Additionally, at stopping time τ we switch from impulse control problem a to impulse control problem p with associated starting value $Z_\tau^p \in \mathcal{Z}^p$. Therefore, $\tau < \infty$ is a stopping time with respect to the filtration $\{\mathcal{F}_t\}$ generated by the Brownian motion $\{B_t\} \in \mathbb{R}^d$. Moreover, note that the environmental stochastic process U_t^p is defined on $\mathbb{R}^m$ and may be different from the process U_t^a; however, the d-dimensional driving Brownian motion B_t is assumed to be the same for both processes.

Let $\mathcal{W}$ denote the set of all admissible combined controls consisting of the impulse control strategy w^a of the τ-stopped impulse control problem a, a finite stopping time τ with associated starting value Z_τ^p for problem p, and the impulse control strategy w^p of the impulse control problem p starting at time τ. This is written as

$$
\begin{aligned}
w & := \ (\theta_1^a, \ldots, \theta_k^a, \tau, \theta_1^p, \ldots; \zeta_1^a, \ldots, \zeta_k^a, Z_\tau^p, \zeta_1^p, \ldots) \qquad (3.81)\\
& = \ (w^a | \tau, Z_\tau^p | w^p) \qquad \in \qquad \mathcal{W}^a \cup \mathcal{W}^\tau \cup \mathcal{W}^p =: \mathcal{W}
\end{aligned}
$$

where k is defined as

$$
k \ := \ \sup \left\{ j \in \mathbb{N}; \theta_j^a < \tau \right\} < \infty \qquad (3.82)
$$

and is assumed finite, which will be justified later. If $w \in \mathcal{W}$ is applied to the system, it takes the form

$$
X_t^a \ = \ X_t^{(w^a)} = \begin{pmatrix} t \\ U_t^a \\ \zeta_j^a \end{pmatrix} \qquad \text{if } \theta_j^a \le t < \theta_{j+1}^a \wedge \tau \quad (j = 0, \ldots, k) \quad (3.83)
$$

or

$$
X_t^p \ = \ X_t^{(w^p)} = \begin{pmatrix} t \\ U_t^p \\ \zeta_j^p \end{pmatrix} \qquad \text{if } \theta_j^p \le t < \theta_{j+1}^p \quad (j = 0, 1, 2, \ldots). \quad (3.84)
$$

We define the profit flow functions $f^a(x^a)$ and $f^p(x^p)$ when the system is in state x^a and x^p, respectively. Assume again that

$$
\mathbb{E}^x \left[\int_s^\tau |f^a(X_t^a)| \, dt + \int_\tau^\infty |f^p(X_t^p)| \, dt \right] < \infty \quad \forall x^a \text{ and } w \in \mathcal{W}. \quad (3.85)
$$

Furthermore, define the switching cost function H^a, H^τ and H^p as

$$
\begin{aligned}
H^a & : \ \mathbb{R}^{n+1} \times \mathcal{Z}^a \times \mathcal{Z}^a \to \mathbb{R}^+, & (3.86)\\
H^\tau & : \ \mathbb{R}^{n+1} \times \mathcal{Z}^a \times \mathcal{Z}^p \to \mathbb{R}^+, & (3.87)\\
H^p & : \ \mathbb{R}^{m+1} \times \mathcal{Z}^p \times \mathcal{Z}^p \to \mathbb{R}^+, & (3.88)
\end{aligned}
$$

strictly positive.

We are now able to formulate the optimization problem of combined impulse control and optimal stopping.

Combined Impulse Control and Optimal Stopping Problem

Find for all $x^a = (s, u^a, z^a)$,

$$\tilde{\phi}^a(x) := \sup_{w \in \mathcal{W}} J^w(x) \qquad (3.89)$$

$$= \sup_{w \in \mathcal{W}} \mathbb{E}^x \left[\int_s^\tau f^a(X_t^a)\, dt - \sum_{j=1}^k H^a \left(X_{\theta_j^a-}^a, \zeta_j^a \right) \right.$$

$$- H^\tau \left(X_{\tau-}^a, Z_\tau^p \right)$$

$$\left. + \int_\tau^\infty f^p(X_t^p)\, dt - \sum_{j=1}^\infty H^p \left(X_{\theta_j^p-}^p, \zeta_j^p \right) \Big| \mathcal{F}_s \right]$$

and find an optimal combined control $\tilde{w} \in \mathcal{W}$ such that

$$\tilde{\phi}^a(x) = J^{\tilde{w}}(x). \qquad (3.90)$$

To facilitate the problem one can see that the last two summands of (3.89) are $\mathcal{F}_\tau$-measurable. Thus, we can rewrite (3.89) as

$$\tilde{\phi}^a(x) = \sup_{w^a \cup w^\tau} \mathbb{E}^{x^a} \left[\int_s^\tau f^a(X_t^a)\, dt - \sum_{j=1}^k H^a \left(X_{\theta_j^a-}^a, \zeta_j^a \right) \right. \qquad (3.91)$$

$$- H^\tau \left(X_{\tau-}^a, Z_\tau^p \right)$$

$$\left. + \underbrace{\sup_{w^p} \mathbb{E}^{x^p} \left[\int_\tau^\infty f^p(X_t^p)\, dt - \sum_{j=1}^\infty H^p \left(X_{\theta_j^p-}^p, \zeta_j^p \right) \Big| \mathcal{F}_\tau \right]}_{(**)} \Big| \mathcal{F}_s \right].$$

$(**)$ is already given by the solution to the impulse control problem (3.40) and (3.41) of Section 3.3 with $\tilde{\phi}^p(X_\tau^p)$ with starting value $x^p = X_\tau^p = (\tau, U_\tau^p, Z_\tau^p)$ where $U_\tau^p = U_{\tau-}^a + \gamma^\tau(X_{\tau-}^a, Z_\tau^{p,d})$ and optimal impulse control $\tilde{w}^p = (\tilde{\theta}_1^p, \tilde{\theta}_2^p, \ldots; \tilde{\zeta}_1^p, \tilde{\zeta}_2^p, \ldots)$. Hence, the optimization problem simplifies to:

Modified Combined Impulse Control and Optimal Stopping Problem

Find for all $x^a = (s, u^a, z^a)$ the function

$$\tilde{\phi}^a(x^a) := \sup_{w^a \cup w^\tau} J^w(x^a) \qquad (3.92)$$

$$= \sup_{w^a \cup w^\tau} \mathbb{E}^x \left[\int_s^\tau f^a(X_t^a)\, dt - \sum_{j=1}^k H^a \left(X_{\theta_j^a-}^a, \zeta_j^a \right) \right.$$

$$\left. - H^\tau \left(X_{\tau-}^a, Z_\tau^p \right) + \tilde{\phi}^p(X_\tau^p) \Big| \mathcal{F}_s \right]$$

and find an optimal combined control $\tilde{w} \in \mathcal{W}^a \cup \mathcal{W}^\tau$ such that

$$\tilde{\phi}^a(x^a) \;=\; J^{\tilde{w}}(x^a). \tag{3.93}$$

3.4.2 Combined Verification Theorem

In order to solve the modified combined control problem in (3.92) and (3.93), we will proceed in a way paralleling the arguments of the impulse control model of Section 3.3. First, we will state conditions when interventions in the impulse control problem a or the stopping problem at time τ take place. This will give us the definitions of the continuation regions for which the system evolves freely without any impulse control or stopping. Then we will again use the notion of stochastically C^2-functions enabling us to state and prove a verification theorem for the combined impulse control and stopping problem.

Define the two maximum operators[44]

$$\mathcal{M}^a\phi(s,u,z) = \max_{\zeta \in \mathcal{Z}^a \setminus \{z\}} \left\{ \phi(s, u + \gamma^a(x,\zeta^d), \zeta) - H^a(x,\zeta) \right\}, \tag{3.94}$$

$$\mathcal{M}^\tau\phi(s,u,z) = \max_{\zeta^p \in \mathcal{Z}^p} \left\{ \phi(s, u + \gamma^\tau(x,\zeta^{p,d}), \zeta^p) - H^\tau(x,\zeta^p) \right\}, \tag{3.95}$$

for all $x = (s,u,z)$ and Borel functions $\phi : \mathbb{R}^{n+1} \times \mathcal{Z}^a \to \mathbb{R}$.

This essentially means that when an impulse control is triggered it will be chosen according to $\mathcal{M}^a\phi$ in (3.94), i.e., the system attains the new state which maximizes its value function. On the other hand, if switching to impulse control problem p, i.e., stopping, is triggered in (3.95), the optimal starting point for problem p that maximizes the value function will be attained. In order to determine which of the two cases (impulse control of a or stopping and switching to p) takes place we define the continuation regions for the two cases as follows:

$$\mathcal{C}^a \;:=\; \left\{ x \in \mathbb{R}^{n+1} \times \mathcal{Z}^a; \; \phi(x) > \mathcal{M}^a\phi(x) \right\}, \tag{3.96}$$

$$\mathcal{C}^\tau \;:=\; \left\{ x \in \mathbb{R}^{n+1} \times \mathcal{Z}^a; \; \phi(x) > \mathcal{M}^\tau\phi(x) \right\}, \tag{3.97}$$

where $\mathcal{C}^a$ is the continuation region of impulse control problem a and $\mathcal{C}^\tau$ is the continuation region of the stopping problem at τ. $\mathcal{C}^a$ and $\mathcal{C}^\tau$ are bounded open sets in $\mathbb{R}^{n+1}$ with boundaries $\partial\mathcal{C}^a$ and $\partial\mathcal{C}^\tau$, respectively.

In order for the stopping problem to make sense, we have to assume that the boundary $\partial\mathcal{C}^\tau$ for which stopping is triggered can be attained by all starting points $X_0 = (s,u,z)$. Therefore, we assume $\mathcal{C}^a \cap \mathcal{C}^\tau \neq \emptyset$ which is a necessary condition for τ to be finite.[45] If this condition did not hold, stopping would never

---

[44]From now on superscripts a for the first impulse control problem are omitted whenever possible.

[45]A somewhat stronger assumption would have been $\mathbb{E}(\tau) < \infty$.

be optimal because the barrier ∂C^a of the impulse control problem would always be hit first bringing the process X^a_{t-} back into the interior of $C^a \cap C^\tau$. The case that an impulse control brings the system in a state X^a_t where stopping is triggered, i.e. $X^a_t \in C^a \cap \bar{C}^\tau$, can not occur due to the triangular inequality we assumed to be satisfied by the function H^a and H^τ.[46] On the other hand this implies that we have to distinguish the following two cases:

1. X^a_{t-} hits first $\partial C^a \Rightarrow X^a_t \in C^a \cap C^\tau$.

2. X^a_{t-} hits first $\partial C^\tau \Rightarrow X^a_{t-}$ is stopped and $X^p_t \in C^p$ due to the triangular inequality of H^τ and H^p.

Example: Consider a perpetual American call on a non-dividend paying asset. From option pricing theory it is known[47] that it is never optimal to exercise the perpetual American call without dividends. Translated to the definition of continuation regions that means that $\tau \to \infty$ and ∂C^τ is not attainable.

Let again Y_t denote the state of the system in time intervals where no impulse control and no stopping takes place. Thus, for fixed $z \in Z^a$ we have the $(n+1)$-dimensional process

$$Y_t = (t, U_t, z) = (t, U_t), \quad t \in [\theta_j, \theta_{j+1}), \quad (j = 0, \ldots, k-1), \quad \text{or}$$
$$t \in [\theta_k, \tau). \tag{3.98}$$

Then Y_t is an Itô diffusion with generator

$$\mathcal{A} = \frac{\partial}{\partial s} + \sum_{i=1}^{n} b_i \frac{\partial}{\partial u_i} + \frac{1}{2} \sum_{i,j=1}^{n} (\sigma \sigma^T)_{i,j} \frac{\partial^2}{\partial u_i \partial u_j} \tag{3.99}$$

with σ^T the transpose of matrix σ and $(\cdot)_{ij}$ $(i, j = 1, \ldots, n)$ the (i, j)th element of the $n \times n$-matrix $\sigma \sigma^T$.

Using again the concept of stochastically C^2-functions we are now able to state the main result of the section.

Combined Impulse Control and Stopping Verification Theorem
(1) Suppose there exists a continuous function $\phi(s, u, z)$ in $\mathbb{R}^{n+1}$ satisfying the

---

[46]Compare the remark to right continuity of the process X_t in connection to the switching cost function H in Section 3.3.
[47]See Karatzas/Shreve [109], p. 60.

following properties:

    (i) ϕ is stochastically C^2 w.r.t. Y_t in $C^a \cap C^\tau$,   (3.100)

    (ii) $\{\phi(\vartheta, U_\vartheta, z)\}_{\vartheta \in \mathcal{T}}$ is uniformly integrable w.r.t. $\mathbb{P}^y$ (3.101)
             for all $z \in \mathcal{Z}^a$ and $\vartheta \in \mathcal{T}$ where $\mathcal{T}$ is the set of all
             $\mathcal{F}_t$-stopping times; $\vartheta < \tau$,

    (iii) $\phi \geq \mathcal{M}^a \phi$ and $\phi \geq \mathcal{M}^\tau \phi$ everywhere,   (3.102)

    (iv) $\mathcal{A}\phi + f \leq 0$ almost everywhere w.r.t. the Green   (3.103)
             measure $G^y(\cdot)$.

Then

$$\phi(x) \quad \geq \quad J^w(x) \tag{3.104}$$

for all $w = w^a \cup w^\tau \in \mathcal{W}$ and all $x = (s, u, z)$.

    (2) Suppose that in addition

$$\mathcal{A}\hat{\phi} + f = 0 \tag{3.105}$$

holds for all $x = (s, u, z) \in C^a \cap C^\tau$. Then define the combined control strategy

$$\hat{w} = (\hat{\theta}_1, \hat{\theta}_2, \ldots, \hat{\theta}_k, \hat{\tau}; \hat{\zeta}_1, \hat{\zeta}_2, \ldots, \hat{\zeta}_k, \hat{Z}_\tau^p) \tag{3.106}$$

inductively as follows: Put $\hat{\theta}_0 = 0$ and for $j = 0, 1, \ldots, k - 1$,

$$\hat{\theta}_{j+1} = \inf\left\{t > \hat{\theta}_j; \ X_t^{(j)} \notin C^a \text{ and } X_t^{(j)} \in C^\tau\right\}, \tag{3.107}$$

$$\hat{\tau} = \inf\left\{t > \hat{\theta}_k; \ X_t^{(k)} \in C^a \text{ and } X_t^{(k)} \notin C^\tau\right\} \tag{3.108}$$

where $X_t^{(j)}$ is the result of applying the control

$$(\hat{\theta}_1, \hat{\theta}_2, \ldots, \hat{\theta}_j, \infty; \hat{\zeta}_1, \hat{\zeta}_2, \ldots, \hat{\zeta}_j, \cdots) \tag{3.109}$$

to X_t. Now choose $\hat{\zeta}_{j+1}$ such that

$$\mathcal{M}^a \hat{\phi}\left(\hat{\theta}_{j+1}, U_{\hat{\theta}_{j+1}}, \hat{\zeta}_{j+1}\right) \tag{3.110}$$

$$= \hat{\phi}\left(\hat{\theta}_{j+1}, U_{\hat{\theta}_{j+1}^-} + \gamma^a(X_{\hat{\theta}_{j+1}^-}, \hat{\zeta}_{j+1}^d), \hat{\zeta}_{j+1}\right) - H^a\left(X_{\hat{\theta}_{j+1}^-}, \hat{\zeta}_{j+1}\right).$$

and for $j = k < \infty$ choose $\hat{Z}_{\hat{\tau}}^p$ such that

$$\mathcal{M}^\tau \hat{\phi}\left(\hat{\tau}, U_{\hat{\tau}}, \hat{Z}_{\hat{\tau}}^p\right) \tag{3.111}$$

$$= \hat{\phi}\left(\hat{\tau}, U_{\hat{\tau}^-} + \gamma^\tau(X_{\hat{\tau}^-}, \hat{Z}_{\hat{\tau}}^{p,d}), \hat{Z}_{\hat{\tau}}^p\right) - H^\tau\left(X_{\hat{\tau}^-}, \hat{Z}_{\hat{\tau}}^p\right).$$

Then $\hat{w} = (\hat{\theta}_1, \ldots, \hat{\theta}_k, \hat{\tau}; \hat{\zeta}_1, \ldots, \hat{\zeta}_k, \hat{Z}_{\hat{\tau}}^p) \in \mathcal{W}$ and

$$\hat{\phi}^a = J^{\hat{w}}(x) = \tilde{\phi}^a(x) \tag{3.112}$$

and the optimal combined impulse control and stopping strategy is given by

$$\hat{w} = \tilde{w}. \tag{3.113}$$

Proof of (1):
Using the generalized Dynkin formula applied to $\phi(s, u, z) = \phi(y)$ and letting $s \leq \theta_j \leq \theta_{j+1} \leq \theta_k$ $(j = 1, \ldots, k-1)$ be two stopping times with respect to $\mathcal{F}_s$ we have

$$\mathbb{E}^y \left[\phi(\theta_{j+1}, U_{\theta_{j+1}}, z) | \mathcal{F}_{\theta_j} \right] \tag{3.114}$$
$$= \phi(\theta_j, U_{\theta_j}, z) + \mathbb{E}^y \left[\int_{\theta_j}^{\theta_{j+1}} A\phi(t, U_t, z)\, dt | \mathcal{F}_{\theta_j} \right]$$

and by (3.103) it follows that

$$\phi(\theta_j, U_{\theta_j}, z) \tag{3.115}$$
$$\geq \mathbb{E}^y \left[\int_{\theta_j}^{\theta_{j+1}} f^a(Y_t)\, dt + \phi(\theta_{j+1}, U_{\theta_{j+1}}, z) | \mathcal{F}_{\theta_j} \right].$$

By the same arguments we have for $s \leq \theta_k < \tau$,

$$\mathbb{E}^y \left[\phi(\tau, U_\tau, z) | \mathcal{F}_{\theta_k} \right] \tag{3.116}$$
$$= \phi(\theta_k, U_{\theta_k}, z) + \mathbb{E}^y \left[\int_{\theta_k}^{\tau} A\phi(t, U_t, z)\, dt | \mathcal{F}_{\theta_k} \right]$$

which gives in view of (3.103)

$$\phi(\theta_k, U_{\theta_k}, z) \geq \mathbb{E}^y \left[\int_{\theta_k}^{\tau} f^a(Y_t)\, dt + \phi(\tau, U_\tau, z) | \mathcal{F}_{\theta_k} \right]. \tag{3.117}$$

Next let $w = (\theta_1, \theta_2, \ldots, \theta_k, \tau; \zeta_1, \zeta, \ldots, \zeta_k, Z_\tau^p)$ be a combined control with $\theta_1 > s = \theta_0$ and let $X_t = X_t^{(w)} = (t, U_t, Z_t)$ be the w-controlled process. Since $\phi \geq \mathcal{M}^a \phi$ holds, we get the following inequality for all $j = 1, \ldots, k-1$:

$$\phi(\theta_j, U_{\theta_j}, Z_{\theta_j}) \;\geq\; \mathbb{E}^y\left[\int_{\theta_j}^{\theta_{j+1}} f^a(X_t)\,dt + \phi(\theta_{j+1}, U_{\theta_{j+1}^-}, Z_{\theta_j})|\mathcal{F}_{\theta_j}\right]$$

$$\geq\; \mathbb{E}^y\left[\int_{\theta_j}^{\theta_{j+1}} f^a(X_t)\,dt \right. \tag{3.118}$$

$$+ \phi\Big(\theta_{j+1}, \underbrace{U_{\theta_{j+1}^-} + \gamma^a(X_{\theta_{j+1}^-}, \zeta_{j+1})}_{=\; U_{\theta_{j+1}}}, \zeta_{j+1}\Big)$$

$$\left. - H^a\left(X_{\theta_{j+1}^-}, \zeta_{j+1}\right)|\mathcal{F}_{\theta_j}\right].$$

Furthermore, in case of stopping ($\phi \geq \mathcal{M}^\tau \phi$) must hold and we get

$$\phi(\theta_k, U_{\theta_k}, Z_{\theta_k}) \;\geq\; \mathbb{E}^y\left[\int_{\theta_k}^{\tau} f^a(X_t)\,dt + \phi(\tau, U_{\tau^-}^a, Z_{\theta_k})|\mathcal{F}_{\theta_k}\right] \tag{3.119}$$

$$\geq\; \mathbb{E}^y\left[\int_{\theta_k}^{\tau} f^a(X_t)\,dt + \tilde{\phi}^p\Big(\tau, \underbrace{U_{\tau^-}^a + \gamma^\tau(X_{\tau^-}, Z_\tau^p)}_{=\; U_\tau^p}, Z_\tau^p\Big) \right.$$

$$\left. - H^\tau\left(X_{\tau^-}, Z_\tau^p\right)|\mathcal{F}_{\theta_k}\right].$$

Summing up and taking expectation with respect to $y = (s, u, z)$ we have

$$\phi(y) \;\geq\; \mathbb{E}^y\left[\int_s^\tau f^a(X_t)\,dt - \sum_{j=1}^k H^a(X_{\theta_j^-}, \zeta_j) \right. \tag{3.120}$$

$$\left. + \tilde{\phi}^p(X_\tau^p) - H^\tau(X_{\tau^-}, Z_\tau^p)\right]$$

$$=\; J^w(y)$$

and (1) is proved.

Proof of (2):
The proof of part (2) is fairly easy. Applying the optimal control times $(\hat{\theta}_1, \hat{\theta}_2, \ldots, \hat{\theta}_k, \hat{\tau})$ to (3.120) and noting that we have equality (3.105) on $\mathcal{C}^a \cap \mathcal{C}^\tau$ between two controls or between control k and stopping in τ, we get equality in (3.120):

$$\hat{\phi}(y) \tag{3.121}$$

$$= \mathbb{E}^y \left[\int_s^{\hat{\tau}} f^a(X_t)\, dt - \sum_{j=1}^k H^a(X_{\hat{\theta}_j^-}, \hat{\xi}_j) + \tilde{\phi}^p(X_{\hat{\tau}}^p) - H^{\hat{\tau}}(X_{\hat{\tau}^-}, \hat{Z}_{\hat{\tau}}^p) \right].$$

This shows, however, that

$$\hat{\phi}(x) = J^{\hat{w}}(x) \tag{3.122}$$

and additionally that $\hat{w} \in \mathcal{W}$ is optimal and

$$\hat{\phi}(x) = \sup_{w \in \mathcal{W}} J^w(x) = \tilde{\phi}(x). \tag{3.123}$$

Remarks:

- A closer look at equation (3.121) reveals the option-like characteristics of the combined impulse control and stopping model. The model can be interpreted as an American option to exchange the impulse control problem a with running reward represented by the first two summands on the left-hand side of (3.121) at time τ for the impulse control problem p which is worth $\tilde{\phi}^p$ by paying the exercise price H^τ.

- In case we have no means to control the first or the second system via impulse control and, additionally, we have also no means to choose an action whenever we stop, the problem simply reduces to an optimal stopping problem. A classical example is the valuation of an American put or call option.

- On the other hand, in our very general model of switching (or exchanging) one impulse control problem to another, we have proved the second important building block for our real options model that is able to value complex option interactions.

- The same extensions concerning the state dependence of the action spaces, the time horizon and the underlying stochastic processes discussed in Section 3.3 are valid here as well. Yet, their mathematical presentation would rather complicate notation than clarify the main approach taken here.[48]

---

[48] The only extension that might be cumbersome is the integration of Poisson jumps into the model. It will not be discussed in detail here. An interesting application to optimal stopping can be found in Mordecki [173].

Chapter 4

Valuing Real Options in a Stochastic Control Framework

As already mentioned, a serious drawback of the real option pricing approach is its inadequacy to deal with many different interacting options embedded in one or more projects. Its lack in dealing with such complexity is a serious limitation to the implementation of the real option approach to real world problems and is in the author's opinion one reason for the slow adoption of real option pricing among corporate practitioners.

Therefore, we make use of the results of Chapter 3 where the two basic mathematical building blocks for the valuation of interacting real options were presented. What remains to show is first to build a link to capital markets in order to apply risk neutral valuation to the generalized impulse control model and to the generalized optimal stopping model. This will be done in Section 4.1 where we will concentrate on the market value of the impulse control problem. Second, the interplay of the different building blocks will be discussed in Section 4.2. A simple graphical decomposition method is presented which simplifies valuation of complex real option interactions significantly without the need to tackle the mathematical complexity as we did in Chapter 3. As it will turn out, we are able to handle more or less any classical real option model with the two generic stochastic control models presented in Chapter 3 and the link to capital markets established in Section 4.1. Afterwards some examples from the literature will be presented in terms of the stochastic control framework.

4.1 Equivalence of Stochastic Control and Contingent Claims Analysis

There are essentially two different ways to value contingent claims in perfect capital markets. The first is by means of constructing a portfolio that replicates the value of the contingent claim at any time up to maturity. It is claimed that the fair (market) price of the contingent claim must equal the value of the replicating portfolio since otherwise there would exist an arbitrage opportunity. Therefore, the seller of a contingent claim can use the trading strategy of the replicating portfolio to hedge her risk by investing in the replicating portfolio. Since the overall portfolio consisting of a short position in the contingent claim and a long position in the replicating portfolio is riskless, it only earns the riskfree instantaneous interest rate r. This allows us to determine all trading strategies.[1] that replicate the contingent claim and therefore the value of the contingent claim itself.[2]

The second approach is to use an equivalent martingale measure which is the result of transforming the original probability measure $\mathbb{P}$ in the real world to an equivalent probability measure $\tilde{\mathbb{P}}$ in the risk neutral world that makes discounting using the risk free interest rate r feasible. This is due to the *Fundamental Theorem of Asset Pricing* introduced by Harrison/Pliska [89, 90]. It relates market completeness and the no arbitrage condition to the existence and uniqueness of an equivalent martingale measure. In fact, if there exist one or more equivalent martingale measures, then the market is said to admit no arbitrage. Furthermore, the equivalent martingale measure is unique if and only if markets are complete. Of course, the opposite is true as well: markets are incomplete, i.e., not every contingent claim can be hedged, if and only if there exists more than one equivalent martingale measure.

Both approaches can be used to derive the market value of the impulse control problem of Section 3.3. While the first approach is intuitively more appealing and provides us with the corresponding trading strategy, the second directly relates the market value of the impulse control model to the optimization problem of stochastic control theory. Therefore, both methods are presented here.

4.1.1 Hedging Portfolio and Fundamental Pricing Equation

In this section it will be shown that, under certain assumptions concerning the impulse control problem in Section 3.3, a hedging portfolio can be constructed

---

[1]The trading strategy is unique if and only if the number of primary securities equals the number of sources of uncertainties. If there are redundant securities traded in the market the replicating trading strategy is not unique. If on the other hand there are not enough primary securities that span all market uncertainties, markets are said to be incomplete. See Bingham/Kiesel [26] for a very good introduction to the subject.

[2]This is the approach taken by the Nobel prize winning authors Robert Merton and Myron Scholes in their seminal works on option pricing [27] and [165] including the late Fischer Black.

such that risk neutral evaluation can be applied. The trading strategy to hedge the risk of the generalized switching option is given in explicit form.

To establish the capital market link is a further important step towards using stochastic control models for valuing real options. Critics often pointed out that the stochastic control approach (or dynamic programming approach) lacks a reasonable derivation of the associated discount rate.[3] In fact, risk adjusted discount rates are different for each control strategy chosen. In view of the possibly infinite number of control strategies it is difficult if not even impossible to determine the corresponding discount rate for each strategy. The criticism at this point completely parallels the criticism that has been directed to decision tree analysis as well.[4] On the other hand, constructing a hedging portfolio allows us to build the bridge to capital markets and enables valuation in a Black/Scholes-world using the riskless instantaneous interest rate r as discount factor for all control strategies. The capital market approach to valuation has therefore two particular advantages. First, from a technical point of view it simplifies valuation significantly by enabling risk free discounting without having to take account for risk premiums in the discount rate. Moreover, it disciplines managerial decision making in that decisions are made such as to optimize the market value of a firm or project.[5] However, since capital market valuation can be applied to any contingent claim regardless of whether we are interested in the market value of a financial contract or real option it, can certainly be applied to the stochastic control problems considered here as well, provided the capital market link is appropriately carried out. This result supports the findings in recent literature that stochastic control and contingent claims analysis are not two distinctive methods but rather two sides of the same coin which can be vitally integrated.[6]

In terms of option pricing theory we are interested in the market value of a contingent claim that has the properties of the impulse control model[7] of Section 3.3. We impose the following modelling assumptions:

- The contingent claim has a fixed expiration date ($T < \infty$) after which it becomes worthless.

- The sources of uncertainty U_t are completely exogenous,[8] i.e.,

$$
\begin{aligned}
b(t, U_t, Z_t) &= b(t, U_t), \\
\sigma(t, U_t, Z_t) &= \sigma(t, U_t), \\
\mathcal{Z}^d &= \emptyset.
\end{aligned}
\tag{4.1}
$$

---

[3] See e.g., Dixit/Pindyck [72] p. 120ff.

[4] See e.g., Copeland/Koller/Murrin [60] for a discussion of the failure of decision tree analysis and a numerical example demonstrating its failure.

[5] See Amram/Kulatilaka [7, 6].

[6] See e.g., Knudsen/Meister/Zervos [114].

[7] Whenever the impulse control model is interpreted as a contingent claim it is called a generalized switching option.

[8] If the functions b and σ depended on the state of the decision variables Z, hedging would be much more difficult. In particular, markets would be incomplete.

That is, U_t follows the stochastic differential equation

$$dU_t \;=\; b(t, U_t)\, dt + \sigma(t, U_t)\, dB_t \qquad \text{with } U_0 = u. \qquad (4.2)$$

- The action space is given by the set $\mathcal{Z} \equiv \mathcal{Z}^i$, i.e., there are only indirect controls.

- Let $w \in \mathcal{W}$ be an admissible impulse control strategy applied by the owner of the contingent claim. Then the state of the system can be described by the process X_t where

$$X_t = X_t^{(w)} = \begin{pmatrix} t \\ U_t \\ \zeta_k \end{pmatrix} \quad \text{if } \theta_j \le t \le \theta_{j+1} \quad (j = 0, \dots, k) \qquad (4.3)$$

with starting value $X_0 = x = (s, u, z)$. Due to the finite expiration date T we denote

$$\theta_k = \sup\{\theta_j;\ \theta_j < T\}. \qquad (4.4)$$

- The instantaneous profit flow is given by the function $f(t, U_t, Z_t)$.

- The switching cost function is given by $H(X_{t-}, Z_t) > 0$ for $Z_{t-} \ne Z_t$.

Let us now turn to the assumptions concerning capital markets in order to derive the market price of the above described generalized switching option. These assumptions do not differ in any way from those typically used in real option pricing. We first assume that the securities market is perfect, i.e., frictionless, and complete in order to ensure that trading in K risky assets is sufficient for dynamic spanning of U_t. Furthermore, there exists a riskless asset whose price β follows the process

$$d\beta_t \;=\; r\beta_t\, dt, \qquad \beta_0 > 0, \qquad (4.5)$$

where the instantaneous riskless rate of return per unit time is assumed to be constant in our analysis. Dynamic spanning of U_t then means that we are able to trade in portfolios that completely replicate the payouts of the generalized switching option or, more precisely, mimics the risk structure of the underlying uncertainties U_t. The price process $M = (M^1, \dots, M^n)$ of these n portfolios $(n \ge d)$ follows

$$dM_t \;=\; \mu_M(t, M_t)\, dt + \sigma_M(t, M_t)\, dB_t. \qquad (4.6)$$

Here, the drift vector $\mu_M \in \mathbb{R}^n$ and the variance matrix $\sigma_M \in \mathbb{R}^{n \times d}$ also satisfy the usual growth and Lipschitz conditions of Section 3.3. Trading in securities takes place continuously, frictionless, and without transaction costs. Moreover short sales of securities are permitted.

All these assumptions are somewhat standard in contingent claims analysis[9] and allow us to form a trading strategy in the riskless asset and the n portfolios that replicate the cash flows of the generalized switching option during its lifetime. It is worth noting that the asset reflected by the generalized switching option itself need not be traded in order to derive its market value. It is sufficient for the payoff streams to be replicatable as long as the firm's goal is to maximize its shareholder value. This means that the shareholders are indifferent between investing into the project or the replicating portfolio(s).

Let $V(x)$ denote the market value of the generalized switching option. We assume that the value function V is stochastically C^2 in $\mathbb{R}^{n+1}$ with respect to the uncontrolled process $Y_t = (t, U_t)$ for fixed $z \in \mathcal{Z}$. Y_t is an Ito diffusion with the following operator:

$$\mathcal{A} = \frac{\partial}{\partial s} + \sum_{i=1}^{n} b_i \frac{\partial}{\partial u_i} + \frac{1}{2} \sum_{i,j=1}^{n} (\sigma\sigma^T)_{i,j} \frac{\partial^2}{\partial u_i \partial u_j} \qquad (4.7)$$

with σ^T the transpose of matrix σ and $(\cdot)_{ij}$ $(i, j = 1, \ldots, n)$ the (i, j)th element of the $n \times n$-matrix $\sigma\sigma^T$. Furthermore, let

$$\delta^i(t, U_t, M_t) = \left[\frac{\mu_M^i(t, M_t)}{M_t^i} - \frac{b^i(t, U_t)}{U_t^i} \right] \qquad i = 1, \ldots, n \qquad (4.8)$$

be the difference between the expected rate of return μ_M^i per unit of the ith portfolio replicating the source of uncertainty U_t^i and the actual expected rate of return b^i per unit of the corresponding source of uncertainty, respectively. This dividend-like payout stream is called *equilibrium rate of return shortfall*,[10] since the actual expected rate of return of the process U_t usually does not match the equilibrium expected rate of return of the replicating portfolio.[11]

To replicate the market value of the generalized switching option consider the set Π which denotes the space of all $\mathbb{R}^{n+1}$-valued, predictable, locally bounded processes in $[0, T]$. We call

$$\pi(t) = (\pi^0(t), \pi^1(t)) \in \Pi \qquad t \in [0, T] \qquad (4.9)$$

a trading strategy (or dynamic portfolio process). We assume that the stochastic integrals

$$\int \pi_t^0 \, d\beta_t \qquad \int \pi_t^1 \, dM_t \qquad (4.10)$$

---

[9]See the discussion of the real option pricing assumptions in Section 2.1.3.

[10]See e.g., McDonald/Siegel [158], Brennan/Schwartz [34].

[11]This relationship can be observed regardless of whether the underlying source of uncertainty is traded or not. In case U_t is the price of oil for example, the rate of return shortfall δ can be interpreted as a *convenience yield*, if there are benefits which an inventory of the good can provide. These benefits are solely of advantage to the owner of the oil but not to the holder of a contingent claim on oil. In case of oil the convenience yield δ can be determined, for example, by futures contracts, see Cherian/Patel/Khripko [48] for an in-depth treatment of the subject.

are properly defined on $[0, T]$.[12]

Here, π_t^0 denotes the amount held at time t of the riskless security (bank account), and π_t^1 denotes the vector of the position in the n replicating portfolios of the vector M_t. The components of the vector π_t^1 may assume negative as well as positive values, reflecting the fact that we allow for short sales and assume that the assets are perfectly divisible. The firm can also borrow money from the bank. That means we allow the firm to lose money as long it is optimal to do so. Consequently, π_t^0 may as well be negative.

Furthermore, the instantaneous cash flow $f(X_t)$ at time t generated by the generalized switching option can be seen as a continuous consumption rate in the replicating portfolio. While switching costs of amount $H(X_{\theta_j^-}, \zeta_j)$ are incurred at each intervention time θ_j, the replicating portfolio process has to mimic these dividend-like payments. Note that $f(X_t)$ may take positive as well as negative values which has to be seen as withdrawal or infusion of funds into the portfolio process. Therefore, we get the market value of the generalized switching option that is replicated by the capital market portfolio as follows:

$$
\begin{aligned}
V(t, U_t, Z_t) &\equiv V(t, M_t, Z_t) = \pi_t^0 \beta_t + \pi_t^1 M_t \qquad\qquad (4.11)\\
&= \underbrace{\pi_0^0 \beta_0 + \pi_0^1 M_0}_{\text{initial endowment}} + \int_0^t \pi_s^0 \, d\beta_s + \int_0^t \pi_s^1 \, dM_s \\
&\quad - \int_0^t f(s, U_s, Z_s) \, ds + \sum_{j:\theta_j \le t} H(X_{\theta_j^-}, \zeta_j).
\end{aligned}
$$

Thus, the portfolio is not self-financing. The last two summands represent the cumulative consumption process which rules the withdrawal and infusion of funds into the replicating market portfolio.

On the other hand, for any admissible impulse control strategy $w \in \mathcal{W}$ we have by the generalized Ito formula for semimartingales that allows for jumps in $[0, T]$:[13]

$$
\begin{aligned}
V(t, U_t, Z_t) &\qquad\qquad\qquad\qquad\qquad\qquad\qquad\qquad (4.12)\\
&= V(0, u, z) + \int_0^t \mathcal{A}V(X_s) \, ds + \int_0^t \nabla V(X_s) \, \sigma(s, U_s) \, dB_s \\
&\quad + \sum_{j:\theta_j \le t} \left[V(\theta_j, U_{\theta_j}, Z_{\theta_j}) - V(\theta_j^-, U_{\theta_j^-}, Z_{\theta_{j-1}}) \right]
\end{aligned}
$$

where the operator $\mathcal{A}$ of V is given by (4.7) and the gradient ∇V is defined as

$$
\nabla V(x) = \left[\frac{\partial V(x)}{\partial u^1}, \ldots, \frac{\partial V(x)}{\partial u^n} \right]. \qquad\qquad (4.13)
$$

---

[12]For the necessary and sufficient conditions for a properly defined trading strategy see e.g., Duffie [75].

[13]See Harrison/Selke/Taylor [92] and the references therein.

In addition to what we would expect from applying Itô's lemma, the last summand of (4.12) states the behavior of the value of the generalized switching option in each single intervention point θ_j up to time t where an impulse control is applied.

Using the fact that the initial endowment of the replicating portfolio in (4.11) equals the market value of the generalized switching option, i.e.,

$$\pi_0^0 \beta_0 + \pi_0^1 M_0 = V(0, u, z), \qquad (4.14)$$

and subtracting (4.11) and (4.12) we get:

$$0 = \int_0^t \Big[\underbrace{\mathcal{A}V(X_s) - \pi_s^0 r \beta_s - \pi_s^1 \mu_M(s, M_s) + f(X_s)}_{(3)} \Big] ds \qquad (4.15)$$

$$+ \sum_{j:\theta_j \le t} \Big[\underbrace{V(\theta_j, U_{\theta_j}, Z_{\theta_j}) - V(\theta_j^-, U_{\theta_j^-}, Z_{\theta_{j-1}}) - H(X_{\theta_j^-}, Z_{\theta_j})}_{(2)} \Big]$$

$$+ \int_0^t \Big[\underbrace{\nabla V(X_s)\, \sigma(s, U_s) - \pi_s^1 \, \sigma_M(s, M_s)}_{(1)} \Big] dB_s.$$

Equation (4.15) must be satisfied for all $t \in [0, T]$, $U_t \in \mathbb{R}^n$ and $Z_t \in \mathcal{Z}$. Therefore, the summands (1)–(3) must each be zero. We check:

(1): Since M^i mimics the risk of uncertainty of U^i unitwise (by construction) we must have

$$\frac{(\sigma)_{ij}}{U^i} = \frac{(\sigma_M)_{ij}}{M^i} \qquad \forall i = 1, \ldots, n; \quad j = 1, \ldots, d. \qquad (4.16)$$

Plugged into (1) this is

$$\nabla V(X_s)\, \sigma(s, U_s) - \pi_s^1 \frac{M_s}{U_s} \sigma(s, U_s) = 0 \qquad (4.17)$$

where $\frac{M_s}{U_s}$ is understood as the vector where division is performed elementwise. Solving this vector equation results in

$$\pi_s^1 = \frac{U_s}{M_s} \nabla V(s, U_s, Z_s) = \left[\frac{\partial V(X_s)}{\partial U^1} \frac{U^1}{M^1}, \ldots, \frac{\partial V(X_s)}{\partial U^n} \frac{U^n}{M^n} \right] \qquad (4.18)$$

which is the first part of the hedging strategy.

(3): Using the result of (1) and setting (3) equal to zero yields:

$$\mathcal{A}V(X_s) - \pi_s^0 r \beta_s - \frac{U_s}{M_s} \nabla V(X_s) \mu_M(s, M_s) + f(X_s) = 0. \qquad (4.19)$$

Solving for π_s^0:

$$\pi_s^0 = \frac{\mathcal{A}V(X_s) - \frac{U_s}{M_s}\nabla V(X_s)\mu_M(s, M_s) + f(X_s)}{r\beta_s}. \tag{4.20}$$

Using the definition of $\delta(s, U_s, M_s)$ in (4.8) and

$$
\begin{aligned}
V(X_t) &= \pi_t^0\beta_t + \pi_t^1 M_t \tag{4.21}\\
&= \frac{\mathcal{A}V(X_t) - \frac{U_t}{M_t}\nabla V(X_t)\mu_M(t, M_t) + f(X_t)}{r\beta_t}\beta_t \\
&\quad + \frac{U_t}{M_t}\nabla V(X_t)M_t
\end{aligned}
$$

we get the *fundamental pricing equation* for the market value of the generalized switching option

$$\frac{\partial V}{\partial t} + \frac{1}{2}\sum_{i,j=1}^n (\sigma\sigma^T)_{i,j}\frac{\partial^2 V}{\partial U^i \partial U^j} + \sum_{i=1}^n (r - \delta^i)U^i\frac{\partial V}{\partial U^i} - rV + f = 0 \tag{4.22}$$

where the time and state dependence of the variables in (4.22) is omitted. However, note that the only variable depending on M_t is $\delta(t, U_t, M_t)$. Equation (4.22) can be written in the more compact form

$$\mathcal{A}^* V - rV + f = 0 \tag{4.23}$$

with the new differential operator

$$\mathcal{A}^* = \frac{\partial}{\partial t} + \frac{1}{2}\sum_{i,j=1}^n (\sigma\sigma^T)_{i,j}\frac{\partial^2}{\partial U^i \partial U^j} + \sum_{i=1}^n (r - \delta^i)U^i\frac{\partial}{\partial U^i}. \tag{4.24}$$

(2): Let us check what would happen with equation (4.15) if we applied an arbitrary impulse control $w \in \mathcal{W}$. First, there may be switching when it is not optimal to switch, so that

$$V(\theta^-, U_{\theta^-}, Z_{\theta^-}) \geq V(\theta, U_\theta, Z_\theta) - H(X_{\theta^-}, Z_\theta) = \mathcal{M}V. \tag{4.25}$$

Consequently, (4.15) would be ≤ 0 implying that the fundamental pricing equation (4.22) would be an inequality

$$\mathcal{A}^* V - rV + f \leq 0. \tag{4.26}$$

But since this strategy reduces the value of the generalized switching option, the firm can clearly do better by investing in the replicating portfolio rather than in the generalized switching option. Note, that (4.25) and (4.26) are the two conditions stated in the first part of the impulse control theorem of Section 3.3.

On the other hand, if $\mathcal{M}V > V$ for an intervention time θ, then in a perfect capital market other firms would wish to purchase the generalized switching option and immediately apply an impulse control. But since this is an arbitrage opportunity $(\mathcal{M}V_{\theta^-} - V_{\theta^-}) > 0$, we must have for the market value of the generalized switching option that the impulse controls are performed optimally. We therefore have the condition:

$$V(\theta_j, U_{\theta_j}, Z_{\theta_j}) - V(\theta_j^-, U_{\theta_j^-}, Z_{\theta_j^-}) - H(X_{\theta_j^-}, Z_{\theta_j}) = 0 \quad \forall j : \theta_j \leq T \ (4.27)$$

which yields that we have equality in (4.15). Furthermore, the fundamental pricing equation (4.22) holds in the continuation region $\mathcal{C}$.

To state the main result, we have found by contingent claim arguments that V satisfies part (1) of the impulse control theorem of Section 3.3 and that $\tilde{V}$ is the market value of the generalized switching option if and only if

$$\mathcal{A}^* \tilde{V} - r\tilde{V} + f = 0 \qquad \text{on } \mathcal{C} \tag{4.28}$$

and

$$\mathcal{M}\tilde{V}(\tilde{\theta}_j^-, U_{\tilde{\theta}_j^-}, \tilde{\zeta}_{j-1}) = \tilde{V}(\tilde{\theta}_j, U_{\tilde{\theta}_j}, \tilde{\zeta}_j) - H(X_{\tilde{\theta}_j^-}, \tilde{\zeta}_j), \ \forall j : \tilde{\theta}_j \leq T. \tag{4.29}$$

Since (4.28) is the condition of impulse control theorem part (2), it follows that the market value $\tilde{V}$ is equivalent to the solution of the corresponding impulse control problem and $\tilde{w}$ is the optimal impulse control strategy.

The above result is of great importance for the valuation of complex real options. From a capital market point of view the generalized switching option is simply a contingent claim whose market value can be determined. From a stochastic control point of view the value of the generalized switching option is the solution to a stochastic optimization problem. The above result shows that using contingent claims analysis or stochastic control leads to the same results when the capital market link is properly established. They can therefore be treated as two sides of the same coin.

So far, the equivalent optimization problem to which the market value $\tilde{V}$ is the solution was not given. Note, that in using contingent claims analysis the hedging and no arbitrage argument replaced the expectation operator and the maximization in the original stochastic control problem. As for the stochastic control problem, we were able to prove that the market value of the contingent claim needs to satisfy the same set of sufficient quasi-variational inequalities as the solution to the stochastic control problem. However, this was achieved using completely different arguments. What remains to show is how to derive the equivalent impulse control optimization problem for which its solution and the market value of the generalized switching option coincide. This is done using the *martingale approach*.

4.1.2   Equivalent Martingale Measure

While we constructed a hedging portfolio which replicates the payoff stream of
the impulse control problem, we were not able to state the optimization problem
to which $\tilde{V}$ is the solution. In order to prove the claim that the market value $\tilde{V}$ of
the generalized switching option and the value function of the equivalent impulse
control problem $\tilde{\phi}$ — which is yet to be determined — coincide, we make use
of the martingale property of the appropriately discounted state price process of
the underlying uncertainties in complete capital markets. This is a very elegant
method which gives further insights into the link between capital market theory
and stochastic control.

Theorem:
The market value of the generalized switching option $\tilde{V}$ coincides with the solution
of $\tilde{\phi}$ of the *equivalent impulse control optimization problem*

$$\tilde{\phi}(x) = \sup_{w \in \mathcal{W}} \tilde{\mathbb{E}}^x \left[\int_0^T e^{-rt} f\left(X_t^{(w)}\right) dt - \sum_{j:\theta_j \leq T} e^{-r\theta_j} H\left(X_{\theta_j^-}, \zeta_j\right) \right] \quad (4.30)$$

for all x where $\tilde{\mathbb{E}}^x$ denotes expectation with respect to the *equivalent martingale
measure* $\tilde{\mathbb{P}}^x$.

Proof:
In fact, this is a direct consequence of the existence and uniqueness of an equivalent
martingale measure $\tilde{\mathbb{P}}$. It implies that for any security with a cumulative dividend
process D and price process S, the sum of the discounted dividends and price

$$\int_0^t e^{-rs} dD_s + e^{-rt} S_t \quad (4.31)$$

is a martingale. That is, given an equivalent martingale measure, the price of the
security can be expressed as[14]

$$S_t = \tilde{\mathbb{E}} \left[\int_t^T e^{-r(s-t)} dD_s + e^{-r(T-t)} S_T | \mathcal{F}_t \right]. \quad (4.32)$$

Consequently, for any arbitrary impulse control $w \in \mathcal{W}$ we should have

$$\begin{aligned}
\phi(x) &= \tilde{\mathbb{E}}^x \left[\int_0^T e^{-rt} f\left(X_t^{(w)}\right) dt - \sum_{j:\theta_j \leq T} e^{-r\theta_j} H\left(X_{\theta_j^-}, \zeta_j\right) \right] \quad (4.33) \\
&= J^w(x) \quad \forall x.
\end{aligned}$$

---

[14]See Duffie [75].

But we have already proved in Section 3.3 that

$$\phi(x) \geq J^w(x) \qquad \forall x \text{ and } w \in \mathcal{W}. \tag{4.34}$$

It follows that $\phi(x)$ is a supermartingale. By part (2) of the impulse control verification theorem of Section 3.3 the equality in (4.34) is attained at

$$\tilde{\phi}(x) = J^{\tilde{w}}(x) = \sup_{w \in \mathcal{W}} J^w(x) \tag{4.35}$$

so that only the solution to the impulse control optimization problem admits the martingale property. This proves that

$$\tilde{\phi}(x) \quad \equiv \quad \tilde{V}(x) \tag{4.36}$$

is the solution to the equivalent impulse control problem as well as the market value (contingent claim value) of the generalized switching option.

What remains to show is the existence and uniqueness of the equivalent martingale measure. Afterwards we will check that $\tilde{\phi}$ satisfies the same system of quasi-variational inequalities as $\tilde{V}$ in (4.25) and (4.26) which underlines the result in (4.36).

Given the assumptions concerning the capital market as above we construct the equivalent martingale measure $\tilde{\mathbb{P}}$ under which the discounted price processes of $M(t)$ are martingales. Then we also get the SDE which $U(t)$ has to satisfy under $\tilde{\mathbb{P}}$.

Remember that the price processes $M_i(t)$ $(i = 1, \ldots, n)$ of the assets spanning the risk of $U_i(t)$ are given by

$$dM_i(t) = \mu_i(t, M(t))dt + \sum_{j=1}^{d} \sigma_{ij}^M(t, M(t))dB_j(t) \quad i = 1, \ldots, n \tag{4.37}$$

and the state price deflator by

$$d\beta(t) \quad = \quad r\beta(t)dt, \qquad \beta_0 > 0, \tag{4.38}$$

assuming for notational convenience r constant. Furthermore, the rate of return shortfall is

$$\delta_i(t, U(t), M(t)) = \left[\frac{\mu_i(t, M(t))}{M_i(t)} - \frac{b_i(t, U(t))}{U_i(t)} \right]. \tag{4.39}$$

In order to find the equivalent martingale measure $\tilde{\mathbb{P}}$, we start with the spanning assets $M(t)$. For the n-dimensional discounted process $M(t)/\beta(t)$ to satisfy the martingale property we must have

$$\frac{M_i(t)}{\beta(t)} = \tilde{\mathbb{E}}\left[\frac{M_i(T)}{\beta(T)} \,\big|\, \mathcal{F}_t \right] \qquad 0 \leq t \leq T, \quad i = 1, \ldots n, \tag{4.40}$$

or equivalently

$$\tilde{\mathbb{E}}\left[d\left(\frac{M_i(t)}{\beta(t)}\right)|\mathcal{F}_t\right] = 0 \tag{4.41}$$

with

$$\beta(t) = e^{rt} \tag{4.42}$$

because the expected returns of the discounted assets in the risk neutral world must equal zero.

Keeping this in mind, the discounted asset price process can be expressed as

$$d\left(\frac{M_i(t)}{\beta(t)}\right) = \frac{1}{\beta(t)}[-rM_i(t)dt + dM_i(t)] \tag{4.43}$$

$$= \frac{1}{\beta(t)}\left[(\mu_i(t, M(t)) - rM_i(t))dt + \sum_{j=1}^{d}\sigma_{ij}^M(t, M(t))dB_i(t)\right]$$

$$= \frac{M_i(t)}{\beta(t)}\left[\frac{\mu_i(t, M(t)) - rM_i(t)}{M_i(t)}dt + \sum_{j=1}^{d}\frac{\sigma_{ij}^M(t, M(t))}{M_i(t)}dB_j(t)\right].$$

In terms of vectors and matrices we can write the differential of the n-dimensional process as

$$d\left(\overbrace{\frac{M_t}{\beta_t}}^{n\times 1}\right) = \frac{1}{\beta_t}[-rM_t dt + dM_t] \tag{4.44}$$

$$= \frac{1}{\beta_t}\left[\overbrace{(\mu(t, M_t) - rM_t)}^{n\times 1} dt + \overbrace{\sigma_M(t, M_t)}^{n\times d}\overbrace{dB_t}^{d\times 1}\right].$$

Omitting the state and time dependence of the variables and noting that the $(d \times d)$-matrix $(\sigma^T\sigma)$ has rank d, because we have assumed market completeness, we get

$$d\left(\frac{M_t}{\beta_t}\right) = \frac{1}{\beta}\left[(\mu - rM)dt + \sigma_M dB_t\right] \tag{4.45}$$

$$= \frac{1}{\beta}\sigma\left[\underbrace{(\sigma^T\sigma)^{-1}\sigma^T(\mu - rM)}_{\vartheta} dt + dB_t\right].$$

Because the d-dimensional process $\vartheta = \vartheta(t, M_t) = (\vartheta_1(t, M_t), \ldots, \vartheta_d(t, M_t))$ given by

$$\vartheta(t, M_t) = \left(\sigma_M^T\sigma_M\right)^{-1}\sigma_M^T(\mu(t, M_t) - rM_t), \quad 0 \le t \le T \tag{4.46}$$

is adapted to the filtration $\mathcal{F}_t$ generated by the d-dimensional Brownian motion B_t we can apply *Girsanov's Theorem*[15] which states that if we define for $0 \leq t \leq T$,

$$\tilde{B}_j(t) = \int_0^t \vartheta_j(s, M_s) \, ds + B_j(t) \qquad j = 1, \ldots, d, \tag{4.47}$$

$$R(t) = \exp\left\{ -\int_0^t \vartheta(s, M_s) \, dB(s) - \frac{1}{2} \int_0^t \|\vartheta(s, M_s)\|^2 \, ds \right\}, \tag{4.48}$$

$$\tilde{\mathbb{P}}(A) = \int_A R(T) \, d\mathbb{P}, \tag{4.49}$$

then, under $\tilde{\mathbb{P}}$, the transformed process

$$\tilde{B}(t) = (\tilde{B}_1(t), \ldots, \tilde{B}_d(t)) \qquad 0 \leq t \leq T \tag{4.50}$$

is a d-dimensional Brownian motion.

$R(t)$ is called the *Radon-Nikodym derivative*.[16] It relates the probability measure $\mathbb{P}$ in the real world to the equivalent probability measure $\tilde{\mathbb{P}}$ in the risk neutral world.[17] Note, that by the existence and the uniqueness of the process ϑ in (4.46) the existence and uniqueness of the equivalent martingale measure $\tilde{\mathbb{P}}$ follows directly from Girsanov's theorem. This completes the sketch of the proof that $\tilde{V} \equiv \tilde{\phi}$.

We now turn back to (4.45) in order to perform the change of measure in the processes M_t and U_t. Therefore define the new Brownian motion

$$d\tilde{B}(t) = \vartheta(t, M_t) dt + dB(t) \tag{4.51}$$

which in fact yields that

$$\tilde{\mathbb{E}}\left[d\left(\frac{M(t)}{\beta(t)} \right) \Big| \mathcal{F}_t \right] = \tilde{\mathbb{E}}\left[\frac{1}{\beta(t)} \sigma_M(t, M(t)) \, d\tilde{B}(t) \Big| \mathcal{F}_t \right] = \vec{0} \tag{4.52}$$

i.e., the discounted asset price processes are martingales under $\tilde{\mathbb{P}}$.

Moreover, the undiscounted process M_t satisfies the following stochastic differential equation (SDE) under $\tilde{\mathbb{P}}$:

$$\begin{aligned} dM_t &= \mu(t, M_t) \, dt + \sigma_M(t, M_t) \, dB_t \tag{4.53} \\ &= \mu(t, M_t) \, dt - \sigma_M(t, M_t) \, \vartheta(t, M_t) \, dt + \sigma_M(t, M_t) \, d\tilde{B}_t \\ &= \mu(t, M_t) \, dt - (\mu(t, M_t) - rM_t) \, dt + \sigma_M(t, M_t) \, d\tilde{B}_t \\ &= rM_t \, dt + \sigma_M(t, M_t) \, d\tilde{B}_t \end{aligned}$$

where we have used the fact that $\sigma(\sigma^T \sigma)^{-1}\sigma^T = \mathbb{I}$.

---

[15] See Girsanov [82].

[16] Provided the integrals in (4.48) are well defined; see, e.g., Bingham/Kiesel [26], p. 47.

[17] A probability measure $\mathbb{P}_1$ is said to be *equivalent* to another probability measure $\mathbb{P}_2$ if they have the same system of nullset, i.e., if $\mathbb{P}_1(A) = 0$ is equivalent to $\mathbb{P}_2(A) = 0$ for all $A \in \mathcal{F}$, see e.g., Duffie [75].

Analogously, the SDE for the sources of uncertainty U_t that are spanned by M_t can be computed to

$$
\begin{aligned}
\mathrm{d}U_t &= b(t, U_t)\,\mathrm{d}t + \sigma(t, U_t)\,\mathrm{d}B_t && (4.54)\\
&= b(t, U_t)\,\mathrm{d}t - \sigma(t, U_t)\,\vartheta(t, M_t)\,\mathrm{d}t + \sigma(t, U_t)\,\mathrm{d}\tilde{B}_t\\
&= b(t, U_t)\,\mathrm{d}t - \sigma(t, U_t)(\sigma_M^T \sigma_M)^{-1}\sigma_M^T\,(\mu(t, M_t) - rM_t)\,\mathrm{d}t\\
&\quad + \sigma(t, U_t)\,\mathrm{d}\tilde{B}_t\\
&= b(t, U_t)\,\mathrm{d}t - \frac{U_t}{M_t}(\mu(t, M_t) - rM_t)\,\mathrm{d}t + \sigma(t, U_t)\,\mathrm{d}\tilde{B}_t\\
&= (r - \delta(t, U_t, M_t))\, U_t\,\mathrm{d}t + \sigma(t, U_t)\,\mathrm{d}\tilde{B}_t,
\end{aligned}
$$

using the definition of δ, the vector division defined as in (4.17) and the unitwise equivalence of the diffusion parameters in (4.16). It is important to note that the discounted process $\mathrm{d}(U_t/\beta_t)$ is not a martingale because of

$$
\tilde{\mathbb{E}}\left[\mathrm{d}\left(\frac{U(t)}{\beta(t)}\right)\Big|\mathcal{F}_t\right] = -\frac{U(t)}{\beta(t)}\delta(t, U_t, M_t)\mathrm{d}t \neq \vec{0}. \tag{4.55}
$$

This also states that the underlying sources of uncertainty U_t earn a below equilibrium rate of return which confirms why δ is often called *rate of return shortfall*.

Now that the SDE of U_t under $\tilde{\mathbb{P}}$ has been derived, we can easily form the quasi-variational inequalities of the impulse control problem which $\tilde{\phi}$ solves. The generator of the impulse control problem in (4.30) can be found by applying Itô's Lemma to the uncontrolled process $Y_t = (t, U_t, z)$ for fixed $z \in \mathcal{Z}$:

$$
\begin{aligned}
\phi(x) &= \tilde{\mathbb{E}}^x\left[\int_0^T e^{-rt} f(Y_t)\,\mathrm{d}t\right] && (4.56)\\
&= f(x)\,\mathrm{d}t + e^{-r\mathrm{d}t}\,\tilde{\mathbb{E}}^x\left[\int_{0+\mathrm{d}t}^T e^{-rt} f(Y_t)\,\mathrm{d}t\right]\\
&= f(x)\,\mathrm{d}t + e^{-r\mathrm{d}t}\,\phi(0 + \mathrm{d}t, u + \mathrm{d}U_t, z)\\
&= f(x)\,\mathrm{d}t + (1 - r\mathrm{d}t)\,\tilde{\mathbb{E}}^x\left[\phi(x) + \frac{\partial\phi(x)}{\partial t} + \sum_{i=1}^n \frac{\partial\phi(x)}{\partial U_i}\mathrm{d}U_i\right.\\
&\quad\left. + \frac{1}{2}\sum_{i,j=1}^n \frac{\partial^2\phi(x)}{\partial U_i \partial U_j}\mathrm{d}U_i\,\mathrm{d}U_j\right]
\end{aligned}
$$

yielding

$$
\begin{aligned}
\phi(x) \;=\; & f(x)\,dt + (1 - r\,dt)\left[\phi(x) + \frac{\partial\phi(x)}{\partial t}dt\right. \\
& + \sum_{i=1}^{n}\left(r - \delta_i\right)U_i\,\frac{\partial\phi(x)}{\partial U_i}dt \\
& \left. + \frac{1}{2}\sum_{i,j=1}^{n}(\sigma\sigma^T)_{ij}\,\frac{\partial^2\phi(x)}{\partial U_i\partial U_j}\,dt\right]
\end{aligned}
\tag{4.57}
$$

which results as expected in the same fundamental pricing equation as in (4.22):

$$
\frac{\partial\phi}{\partial t} + \frac{1}{2}\sum_{i,j=1}^{n}(\sigma\sigma^T)_{i,j}\,\frac{\partial^2\phi}{\partial U_i\partial U_j} + \sum_{i=1}^{n}\left(r - \delta_i\right)U_i\,\frac{\partial\phi}{\partial U_i} - r\phi + f = 0 \tag{4.58}
$$

$$
\Longleftrightarrow \qquad \mathcal{A}^*\phi - r\phi + f = 0. \tag{4.59}
$$

Therefore the set of quasi-variational inequalities for the impulse control problem (4.30) is given by

$$
\begin{array}{ll}
(1) & \mathcal{A}^*\phi(x) - r\phi(x) + f(x) \le 0, \\
(2) & \phi(x) \ge \mathcal{M}\phi(x), \\
(3) & (\phi(x) - \mathcal{M}\phi(x))(\mathcal{A}^*\phi(x) - r\phi(x) + f(x)) = 0, \\
(4) & \phi(T, U_T, Z_T) = 0,
\end{array}
\tag{4.60}
$$

and $\tilde{\phi}(x) \equiv \tilde{V}(x)$ is the unique solution of (4.60).

4.1.3 Interpretation and Conclusions

The same procedure can, of course, be extended to the generalized timing option. While the corresponding hedging strategy would only slightly change, the equivalent martingale measure would need to be extended to the probability measure of the second impulse control problem.

To conclude, by interpreting the impulse control problem of Section 3.2.2 as a contingent claim, the market value of the generalized switching option has been derived using contingent claims analysis. The connection between stochastic control and the two valuation approaches of contingent claim analysis is summarized in Table 4.1.

As already mentioned, stochastic control problems can be seen as contingent claims in capital market theory whose market value can be derived using capital market valuation techniques. To distinguish the terminology involved we called the contingent claim of an impulse control model generalized switching option and

Stochastic Control	Contingent Claim Analysis	
	Risk Neutral Valuation	Hedging Strategy
Impulse Control	Generalized Switching Option	
Combined Impulse and Stopping	Generalized Timing Option	
Optimization	Optimization	No Arbitrage Condition
Expectation w.r.t. $\mathbb{P}$	Expectation w.r.t. $\tilde{\mathbb{P}}$	Hedging Portfolio
Risk Adjusted Discount Rate ρ	Riskless Discount Rate r	
Subjective Value	Market Value	

Table 4.1: Comparison of Stochastic Control and Contingent Claim Analysis

the combined impulse control and optimal stopping model a generalized timing option. The two contingent claim approaches considered were risk neutral valuation using an equivalent martingale measure and valuation using a hedging strategy. Both approaches were employed to derive the market value of the corresponding impulse control model. The contingent claim approach was necessary to derive market values because stochastic control does not give us a hint which discount rate can be used to find the value function of a stochastic optimization problem. The stochastic control approach starts by specifying a discount rate ρ exogenously as a part of the objective function. Therefore, the objective function can only reflect the decision maker's subjective valuation of risk. In contrast, in the contingent claim approach the required rate of return on the asset was derived as an implication of the overall market equilibrium in capital markets. Only the riskless rate of return r was taken to be exogenous. Thus, the contingent claim approach offers a better treatment of the discount rate if we are interested in market values rather than in subjective individual investor's values.

We were able to derive the market value of the generalized switching option using a hedging strategy and the no arbitrage condition. Furthermore, it was shown that the contingent claim value is in fact the solution to an impulse control problem. Comparing the two approaches the hedging portfolio corresponds to taking expectation in the stochastic control problem. Furthermore, the no arbitrage condition was equivalent to maximizing over all impulse control strategies.

However, the valuation approach using a hedging strategy did not provide us with the equivalent impulse control problem in stochastic control whose solution directly yields the market value of the impulse control problem. To circumvent this drawback we used the notion of equivalent martingale measures which combines the capital market assumptions of the hedging procedure with the optimization of stochastic control justifying risk neutral valuation in the equivalent impulse control problem. Consequently, in order to build the bridge from stochastic control to contingent claims analysis in complete capital markets, one has only to perform a measure transformation from the probability measure $\mathbb{P}$ in the real world to the equivalent martingale measure in the risk neutral world $\tilde{\mathbb{P}}$, which accordingly transforms the original stochastic control problem to an equivalent stochastic control problem whose value function reflects the market value.

The contribution of the combination of stochastic control and capital market valuation is twofold. First of all, it combines the theory of stochastic control with the field of mathematical finance, including the real option approach. These two streams of research are often treated separately without using the extent of their close resemblance. Moreover, the practical relevance of the stochastic control approach should be stressed. Building the bridge to capital markets allows for valuing complex options that have not been developed in mathematical finance and especially in real option theory so far.

On the other hand, the utilization of stochastic control models requires some deeper mathematical knowledge which may cause management to refrain from employing it. However, as will be shown next, a simple and intuitive graphical decomposition method to represent the contingency structure of interacting real options can be developed, which is able to overcome this obstacle. Furthermore, the decomposition method will give rise to efficient computational implementation of arbitrary combinations of the two generic real option models presented above.

4.2 Contingency Structure of Option Interactions

It is claimed here that any decision situation with complex real option interactions can be decomposed into combinations of generalized timing options and generalized switching options. In order to achieve a broader acceptance of the proposed method by management who is often not trained in dealing with complex mathematical problems, a graphical representation method that describes the interplay of different embedded real options is developed. Such a decomposition method should also allow for easy computational implementation.

According to Section 4.1.2 the incorporated real options are already identified and classified. In order to find a graphical representation of the contingency structure and subsequently employ a numerical valuation technique, the following procedure is suggested:

1. Define states of the system.

2. Characterize type of switches between the states.

3. Group states which are costly reversible and describe corresponding generalized switching option.

4. Describe stopping models (generalized timing options).

5. Check whether optimization problems are well posed.

6. Represent decision situation as graphical decomposition.

The possible states of the firm or project to be valued can be derived by a thorough analysis of the decision situation. After having framed the application it should be clear how and when each state can be attained. For example, consider the simplest case of the option to wait to invest.[18] If there is no further flexibility associated with the investment project, there are only two possible states, namely inactive before the investment is triggered and active after the investment has taken place. Another example is the option to temporarily shut down and reopen.[19] There the firm can be either inactive, active or temporarily shut down. Therefore three different states have to be considered.

The next step is to characterize the type of switches between the different identified states. The following cases can be distinguished. As before the switches can be either irreversible or (costly) reversible giving rise to generalized timing options and generalized switching options. In addition, we have to take account of situations where the decision time is a priori fixed, which is the case for any European type real option. Furthermore, the period in which the option can be exercised may be naturally fixed and the option expires afterwards. The former may happen for example when the exercise date is contractually specified. The latter may be the case of a lease giving its holder the right to exercise up to a certain expiration date. Lastly, the question whether the real option depends on subsequent real options or not has to be answered. The different types of switches to be considered are summarized in Figure 4.1.

In the next step states which are costly reversible can be summarized to generalized switching options. In order to implement these generalized switching options as impulse control models, the following information has to be gathered for each impulse control model i:

- the first possible entry time and the latest possible exit time of i,

- the underlying sources of uncertainty U of model i,

- the action space Z between which reversible switches take place,

- the possible starting values U_t and Z_t at time t,

- the equivalent probability measure $\mathbb{P}$,

- the switching cost function H, and

- the instantaneous cash flow function f of the generalized switching option i.

The remaining states that cannot be summarized are called *single states*. Since only irreversible switches lead to exit these kind of states, the only information required is:

---

[18] See the seminal paper of McDonald/Siegel [160] and for recent different mathematical approaches Dixit/Pindyck/Sødal [74] and Sødal [213] who interpret the option to wait as a mark up problem; another approach is provided by Alvarez [5] in terms of Green kernels.

[19] See Dixit/Pindyck [72], p. 229ff.

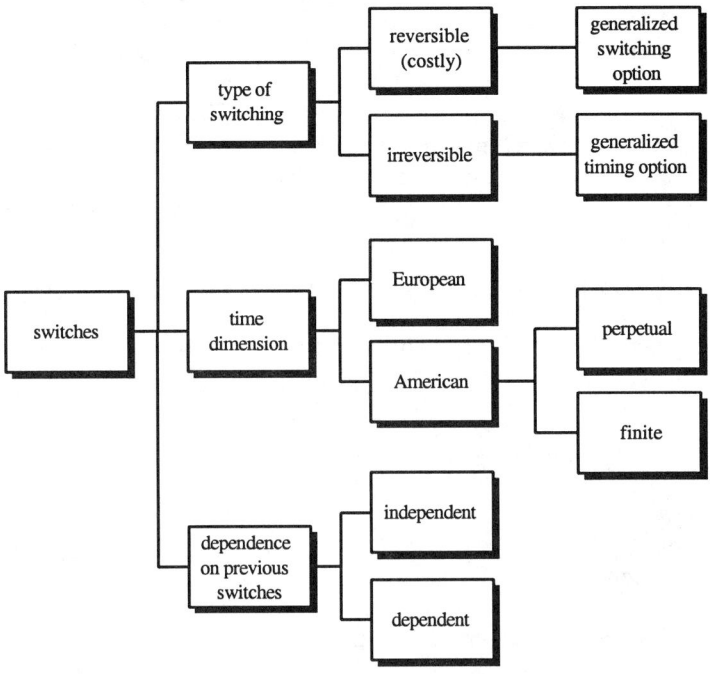

Figure 4.1: Characteristics of switches

- the first possible entry time and the latest possible exit time of state i,

- the underlying sources of uncertainty U of state i,

- the possible starting value U_t at time t,

- the equivalent probability measure $\mathbb{P}$, and

- the instantaneous cash flow function f of state i.

Furthermore, terminal or abandonment values have to be added for each state depending on whether abandonment is possible.[20]

Let us now turn to the characterization of irreversible switches. In addition to the already noted distinction between American and European types in timing

---

[20]Although switching to the state of abandonment is not treated explicitly in this discussion, it is an option that is naturally contained in real life investment projects because disinvestment is always possible even though sometimes not useful. For that reason the abandonment state will be represented by its own graphical modelling element.

of switches, we also have to take into consideration possible control opportunities when switching occurs. This boils down to options which allows one to choose the maximum of several assets. We will call these control opportunities the *intensity of investment*.[21] This extension of the option specification has not been recognized in the classical literature of real options. It will therefore be discussed as an example of the graphical decomposition method later in this chapter. It is important to note that intensity may occur over discrete or continuous control variables. Finally, as input for the option valuation model the following information is needed:

- the first and last possible exercise time of the irreversible switch,

- the number and nature of control opportunities that allow for intensity of investment, and

- the switching cost function H^τ.

With regard to the previous exposition of the impulse control and stopping model, it has to be checked whether the impulse control and optimal stopping models represented by the sequences of reversible and irreversible switches are well posed. That means we have to check if the conditions that have to hold in order to apply the preceding verification theorems are satisfied. Special care has to be taken to the assumption that the value function may not blow up; particularly, if models with infinite time horizon are considered. The question of well posedness has to be answered for each application and cannot be dealt with in general.

After having finished the previous steps the contingency structure of the real option interactions can be represented by means of the graphical decomposition using the graphical modelling elements shown in Figures 4.2 to 4.7.

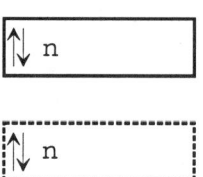

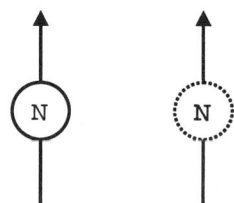

Figure 4.2: Modelling element generalized switching option with n states between which reversible switching is possible. Dashed line denotes currently inactive generalized switching option. Left end of the box represents first possible entry time and the right end the latest possible exit time, respectively.

Figure 4.3: Modelling element irreversible switching where N denotes the number of states over which switching is optimized. Solid circle denotes American type switching — original generalized timing option — and dotted circle its European counterpart.

---

[21] See Bar-Ilan/Strange [15].

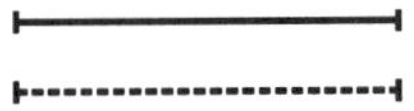

Figure 4.4: Modelling element single state. Dashed line represents currently inactive state. Left end first possible entry time, right end latest possible exit time.

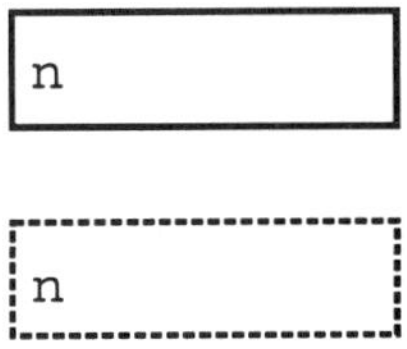

Figure 4.5: Modelling element n parallel states where no switching between states is allowed.

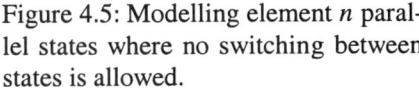

Figure 4.6: Modelling element infinite time horizon.

Figure 4.7: Modelling element abandonment. First case, natural end of system. Second case, abandon for salvage value of the system.

In order to demonstrate how the main graphical modelling elements can be used to decompose complex option structures in a simple graphical manner some examples are given next. We will start with plain financial options before turning to real options.

Consider the simplest case of a European call or put option. The decision to be made is whether the option should be exercised at its expiration date T. There are two states to be considered. During the time interval $[0, T)$ the state is holding the option and the option is not exercised. At time T a distinction must be made between the case where the option is exercised, which leads to switching over to the second state, and where the option is not exercised, which means there is no change in the state of the system. At time T these two states accrue from the payoff function of the option. For example the payoff function of a European call is given by

$$C(T, S_T) = \max\{S_T - K, 0\} \tag{4.61}$$

where S_T is the value of the asset underlying the option at time T and the switching costs H are equal to the exercise price K. Without going too much into the mathematical details, the graphical representation of the option can be found in Figure 4.8.

As can be seen in the graph, during the time interval $[0, T)$ the state of the system remains unchanged because there is no flexibility to alter anything. However, at time T there are two opportunities to take action. If the holder of the option decides to exercise, the system performs a European switch, represented by the

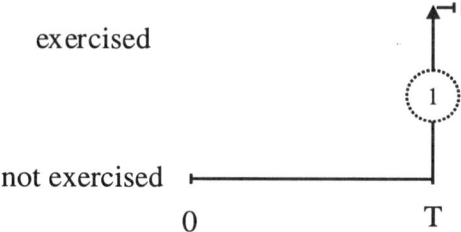

exercised

not exercised

0   T

Figure 4.8: Graphical representation of a European option

dotted circle, and immediately afterwards ends. Since there is only one state over which switching can be chosen, the dotted circle contains 1. If the option is not exercised, the system will not change its state and its natural end is represented by the right closure of the solid line.

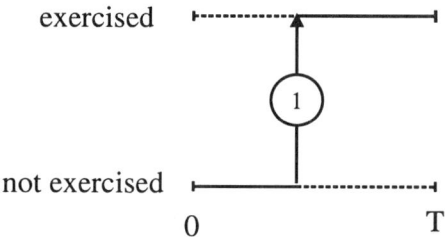

exercised

not exercised

0   T

Figure 4.9: Graphical representation of an American option

The next example deals with the representation of an American option in the graphical decomposition model. In much the same way as for its European counterpart there are two states to consider, see Figure 4.9. The distinctive feature of the American option is that the second state, option exercised, can be attained everywhere in the time interval $[0, T]$ so that its first possible entry time is 0. While the system starts in the not exercised state, it is drawn as a solid line up to the *random* stopping time τ where exercising is triggered. Afterwards the state not exercised is inactive denoted by the dashed line. At time τ the switch to the state exercised is performed which is afterwards active up to time T which would have been the latest possible entry *and* exit time. Since there is only one state to switch to, the solid circle indicating an American switch contains 1. The natural end of both states is time T.

The last example of a financial option is a comparison between an American and European compound option. The American contract allows its holder to exchange the compound option up to time T_1 with a simple option on the underlying expiring at time $T_2 > T_1$. For example, the resulting intrinsic value of an American compound option for two call options with exercise price K_1 is given by

$$\max\{c(t, S_t) - K_1, 0\} \tag{4.62}$$

and the corresponding value of the compound option by

$$cc(t, x) = \sup_{t \le \tau_1 \le T_1} \tilde{\mathbb{E}}^x \left[e^{-r(\tau_1 - t)} \max\{c(\tau_1, S_{\tau_1}) - K_1, 0\} \right] \tag{4.63}$$

where $cc(t, x)$ is the value of the compound option (call on a call), $c(t, x)$ is the value of the underlying call option, and S_t is the call option's underlying source of uncertainty, e.g., the stock price. $\tau_1 \in [0, T_1]$ is the stopping time at which the American compound option is exercised.

In contrast, exercise of the European compound option and its underlying simple option may only take place at the fixed times T_1 and T_2, respectively. Consequently, we get for the European compound option for two calls the value function

$$CC(t, x) = \tilde{\mathbb{E}}^x \left[e^{-r(T_1 - t)} \max\{C(T_1, S_{T_1}) - K_1, 0\} \right] \tag{4.64}$$

reflecting the restriction that the exercise date may not be chosen. The differences in the graphical representation are demonstrated in Figures 4.10 and 4.11.

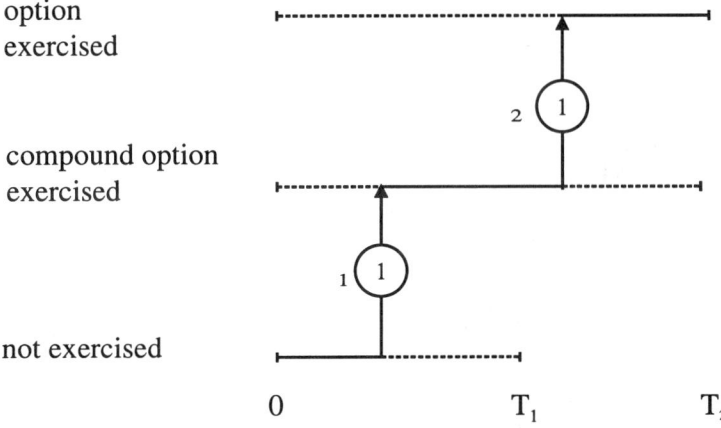

Figure 4.10: Graphical representation of an American compound option

Since the American compound option and its underlying simple option may be exercised instantaneously, the first possible entry time into both states is zero. On the other hand the latest exit time for both states is T_2. In much the same way the time intervals over which the states of the European compound option are

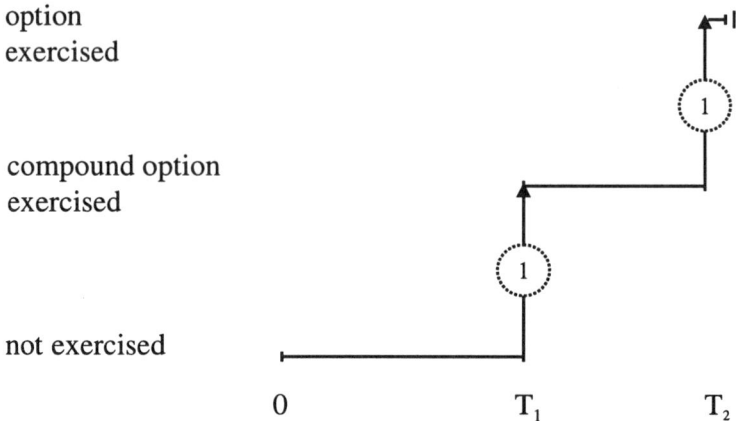

option
exercised

compound option
exercised

not exercised

0   T₁   T₂

Figure 4.11: Graphical representation of a European compound option

active can be obtained. Due to the fixed switching times in the European case the considerations to get the result of Figure 4.11 are straightforward.

Although the previous examples have been devoted to financial options, the same logic of graphical representation applies to real options. How this can be done is demonstrated by two simple real options, namely the perpetual option to invest (or option to wait) which illustrates irreversible switches and the entry and exit decision of firms which highlights the use of costly reversible switches. Afterwards the more complex model of real option interactions introduced by Tirgeorgis [223] will be discussed in the context of the graphical decomposition method proposed here.

The simplest real option that is discussed in the literature is the option to wait. It is simply a perpetual American call on the project value V which can be exercised by paying the investment sum I. The value of this option is given by

$$C(t, x) = \sup_{t \leq \tau < \infty} \tilde{\mathbb{E}}^x \left[e^{-r(\tau - t)} \max\{V_\tau - I, 0\} \right]. \qquad (4.65)$$

Having only two states to consider (inactive vs. active) as for any simple option with no control opportunity when exercising is triggered and the irreversible switch is performed, the graphical decomposition of this simple example is shown in Figure 4.12.

It is important to note that the graphical decomposition method does not give us a clue whether the option is actually exercised. If, for example, for the option to wait the underlying is a stock price, then the perpetual American call would never be exercised because the risk neutral stochastic price process would grow at the same rate as the risk neutral asset, permitting no advantage of holding the stock vs. holding the money in the bank account. If on the other hand the underlying is a real

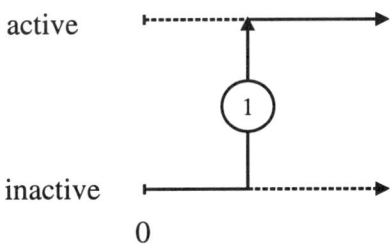

active

inactive

Figure 4.12: Perpetual option to invest

asset earning a below equilibrium rate of return, e.g., a plant outlay, then the option will be exercised with a strictly positive probability as long as the expected risk neutral rate of growth is less than the risk-free interest rate.[22] This result can only be obtained by an in-depth analysis of the mathematical properties of the incorporated real options stressing that the well posedness of the optimization problems has to checked.

An interesting example of an application of the generalized switching option to real options analysis is given in the book of Dixit/Pindyck [72]. They model the entry and exit decision of a firm. Although not mentioning explicitly that entry and exit decisions under uncertainty fall into the framework of impulse control problems, they use the usual smooth pasting and value matching conditions in order to get equations whose solutions yield the parameters of the value function and the trigger levels for — in their case — output prices which signal entry and exit. The main economic result of their model is the ability to explain the well-known hysteresis effect in the investment behavior of firms.[23] The hysteresis effect explains the firm's willingness in the presence of irreversibility and uncertainty to operate in situations where the difference between prices and variable costs is negative, while maintaining the chance that prices will rise in the near future rather than disinvest and exit the market. Vice versa, as long as the firm is inactive it will remain inactive as long as the net present value of the investment is positive enough to weigh out the lumpy investment amount I plus the option to invest later on. This results in an entry trigger price which is larger than the break even price resulting from a naive net present value investment rule. Thus, ceteris paribus, the entry/exit trigger price is larger/smaller than the static NPV trigger price resulting in a band of inertia — the hysteresis band. In essence, the sequential entry and exit decision of a firm can be seen as a sequence of compound investment and abandonment options.

---

[22]See McDonald/Siegel [158] for a discussion and Shreve/Chalasani/Jha [204] for mathematical proofs.

[23]The hysteresis effect is widely discussed in microeconomic literature as part of the investment uncertainty relationship, see e.g., Abel/Eberly [3], Abel/Dixit/Eberly/Pindyck [2], Alvarez [5], Dixit [70].

Formulating this investment problem in terms of the graphical decomposition method, we start by identifying two states — inactive (0) vs. active (1). Switching between the states is costly reversible, so that we have the case of a simple generalized switching option with two possible states. The needed information to formulate the impulse control problem is the time horizon, which is assumed to be infinite, the underlying source of uncertainty, which is the output price P following a geometric Brownian motion, the action space $\mathcal{Z}$ consisting of just the states active and inactive, and the switching cost function H which is given by the matrix

$$H(t, P_t, Z_t, 1 - Z_t) = \begin{pmatrix} 0 & I \\ E & 0 \end{pmatrix} \qquad (4.66)$$

where $I > 0$ is the entry investment cost and $E \neq 0$ is the market exit cost. Moreover, we have the instantaneous cash flow f which is

$$f(t, P_t, Z_t) = \begin{cases} P_t - C \text{ for } Z_t = 1, \\ 0 \quad \text{ for } Z_t = 0, \end{cases} \qquad (4.67)$$

where C denotes the variable cost when the firm is active. There is no stopping option to consider because switching between the two states is reversible. The optimization problem is well posed as long as the conditions of the impulse control verification theorem are met. Since the exogenous output price process follows geometric Brownian motion according to

$$dP_t = (r - \delta) P_t \, dt + \sigma P_t \, d\tilde{B}_t, \qquad P_0 = p \qquad (4.68)$$

under the equivalent martingale measure $\tilde{\mathbb{P}}$, switching takes place only for $\delta > 0$. Otherwise waiting would always be better than investing.[24] Hence, if the assumption that δ is greater than zero is met, the optimization problem is well posed and admits a solution which can easily be shown.

active

inactive

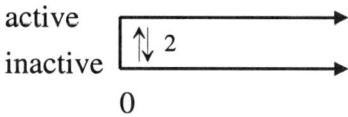

0

Figure 4.13: Optimal entry and exit decision

The graphical representation of the real option model is displayed in Figure 4.13. Since the switching option between the two states is already active at time 0 regardless which initial state — 0 or 1 — is chosen and has infinite time horizon, it is drawn as a solid right open box. The up and down arrows exhibit the opportunity to switch reversibly between the two states.

---

[24] See Dixit/Pindyck [72], p. 186.

So far the considered option interactions are quite simplistic, like most models that have been discussed in the literature so far. Simple decision situations with little incorporated flexibility are interesting from an academic point of view because they sometimes admit closed form solutions and can therefore be analyzed with regard to special topics of interest, as for example the relationship of investment behavior and uncertainty or the comparison of different decision strategies. However, they are far from being able to portray the complex structure of real-world decisions. As the next example demonstrates, the graphical decomposition method enables decision makers to develop models that incorporate many real options in an intuitive and easy way without losing the link to the two generic real options identified above as the generalized timing option and the generalized switching option.

For demonstration purposes we concentrate on the well-known example of option interactions developed by Trigeorgis.[25] It is shown how his model of different real options embedded in the building stage and operating stage during the whole life of an investment project can be decomposed into a sequence of generalized timing options, therefore admitting a representation in the graphical decomposition method.

We start with a description of the decision situation and then work through steps 1 to 6 in order to find the graphical representation of the model. Consider a firm that has the opportunity to invest into a project consisting of a building stage where several investment outlays have to be made at certain times before the operating stage begins and cash flows are generated. For simplicity it is assumed that the static net present value of the project is the only source of randomness. The static NPV follows a geometric Brownian motion according to

$$dV_t = (r - \delta) V_t \, dt + \sigma V_t \, d\tilde{B}_t, \qquad V_0 = v \qquad (4.69)$$

under the equivalent martingale measure $\tilde{\mathbb{P}}$, therefore it is assumed that spanning holds. The investment opportunity allows management the flexibility to[26]

- defer undertaking the project up to T_1 years,

- permanently abandon construction, with no recovery, by forgoing subsequent planned investment outlays,

- contract the scale of the project by reducing planned investment outlays,

- expand the project's scale by making an additional investment outlay, and

- switch the operating project to its best alternative use modelled as a specified salvage value.

After the decision to invest in the project has been made at a certain stopping time $\tau \in [0, T_1]$, an initial outlay I_1 is made. Two more investment outlays — I_2 and

---

[25]See his PhD thesis [220] and the reprints of the same model in [223] and [225], Chapters 7 and 10.
[26]See Trigeorgis [225], p. 228.

I_3 at T_2 and T_3 years after starting the project — are necessary to conclude the construction stage. In exchange for making the investment outlay I_3 the firm may also decide to contract the scale of the project by $c\%$ of the static project value V by investing only $I_3 - \tilde{I}_3$. By entering the operating stage the project starts to generate cash flows. T_4 years after the project was initiated management has the further option to expand the scale of the project by $e\%$ of the static project value V by making an additional investment outlay I_4. For simplicity it is assumed that during the construction stage the project may only be abandoned at the times T_1, T_2 and T_3. In the operating stage the project may be abandoned at any time for its salvage, value which is assumed to be 50% of the cumulative investment outlay depreciated exponentially at $d\%$ per year. The project stops T_5 years after it has been initiated.

Starting with step 1 the states inherent in the decision situation are summarized in Table 4.2. While there is no reversibility between the states, all switches have

No.	States	Earliest Entry	Latest Exit
1	inactive (waiting)	0	T_1
2	first building stage	0	$T_1 + T_2$
3	second building stage	T_2	$T_1 + T_3$
4	operating full	T_3	$T_1 + T_4$
5	operating contracted	T_3	$T_1 + T_4$
6	operating full/expanded	T_4	$T_1 + T_5$
7	operating full/full	T_4	$T_1 + T_5$
8	operating contracted/expanded	T_4	$T_1 + T_5$
9	operating contracted/full	T_4	$T_1 + T_5$
10	abandoned	T_4	$T_1 + T_5$

Table 4.2: State description of Trigeorgis´ option interaction model

to be characterized either as generalized timing options or European type options on the maximum of several assets. Furthermore, the source of uncertainty for all states is just the static NPV V. Since the cash flows of the system in each state are already modelled by the evolution of the static NPV, the instantaneous cash flow function f vanishes.[27] The specification of the investment project gives rise to two generalized stopping options and several European type options. The first gener-

---

[27] It is in general a quite unrealistic assumption to use the static gross project value V as the underlying asset because it implies that there would be investors willing to buy the static NPV while forgoing the advantages of active management of the incorporated real options. However, any rational investor would prefer to invest in the project including the real option value rather than in the replicating portfolio of the static project value. Consequently, this results in an arbitrage opportunity which evades the justification of valuing the incorporated real options using the capital market analogy. For that reason real option models should be oriented according to an explicit modelling of the dependence between the instantaneous cash flow function f and its influencing sources of uncertainty, e.g., output or input prices. However, for expositional purposes we keep the modelling framework of Trigeorgis.

alized stopping option is the option to defer investment up to T_1. This is simply an American call option with finite time horizon on the expanded project value. The second generalized timing option is the ability to sell the operating project through its whole lifetime whenever the salvage value exceeds the value of continuing the project. All other flexibilities can be modelled in terms of European options.

Switch $1 \to 2, 10$ at time τ:
Let $V_i(t)$ be the expanded project value at time t, i.e., including all subsequent real options, if the project is currently in state i and $S_i(t)$ the corresponding abandonment value. Then the irreversible switches between the different states of the project can be characterized in the following way. The option to defer the project up to T_1 is either exercised at time τ or expires worthless at T_1 (state 10 with $V_{10} = 0$).

$$V_1(0) = \sup_{0 \leq \tau \leq T_1} \tilde{\mathbb{E}} \left[e^{-r\tau} \max\{V_2(\tau) - I_1, 0\} \right]. \tag{4.70}$$

Switch $2 \to 3, 10$ at time $\tau + T_2$:
After having decided to invest into the first construction stage (state 2) the next decision to be made at time $\tau + T_2$ is whether to go on to the second construction stage (state 3) or abandon the project (state 10) by forgoing I_2. This is a simple European call option.

$$V_2(\tau + T_2) = \max\{V_3(\tau + T_2) - I_2, 0\}. \tag{4.71}$$

Switch $3 \to 4, 5, 10$ at time $\tau + T_3$:
The last investment outlay at time $\tau + T_3$ after which the operating stage is entered can be either made fully (state 4) by paying I_3, contracted (state 5) by paying $I_3 - \tilde{I}_3$ and receiving only $(1 - c)\%$ of the project value or not at all (state 10). In this case switching is a European option on the maximum of two assets (not regarding the abandonment state) which represents an intensity option in its simplest form.

$$V_3(\tau + T_3) = \max\{V_4(\tau + T_3) - I_3, \ V_5(\tau + T_3) - (I_3 - \tilde{I}_3), \ 0\}. \tag{4.72}$$

Switch $4 \to 7, 6, 10$ at time $\tau + T_4$ or $4 \to 10$ at time τ_A:
If it has been decided to operate at full scale there are two situations to distinguish. First, since the project is now completed, it may be sold for its salvage value S_4 during the time interval $[\tau + T_3, \tau + T_4]$ which is a generalized stopping option (switching to state 10). Alternatively, if the option to sell the project is not exercised until $\tau + T_4$ the project may be either sustained (switching to state 7), expanded by $e\%$ by making the investment outlay I_4 (state 6) or abandoned for its salvage value (switching to state 10).

$$
\begin{aligned}
V_4(\tau + T_3) &= \sup_{\tau + T_3 \leq \tau_A \leq \tau + T_4} \tilde{\mathbb{E}} \left[e^{-r(\tau_A - \tau - T_3)} \max\{V_4(\tau_A), \ S_4(\tau_A)\} \right], \\
V_4(\tau + T_4) &= \max\{V_7(\tau + T_4), \ V_6(\tau + T_4) - I_4, \ S_4(\tau + T_4)\}, \\
S_4(\tau_A) &= 0.5 \cdot \left[I_1 e^{-d(\tau_A)} + I_2 e^{-d(\tau_A - T_2)} + I_3 e^{-d(\tau_A - T_3)} \right].
\end{aligned}
\tag{4.73}
$$

Switch 5 → 9, 8, 10 at time τ + T₄ or 5 → 10 at time τₐ:
The same, of course, is true if the prior decision was to contract the scale of the project. Switching may then take place to state 9 (contracted/full), state 8 (contracted/expanded), or the project may be abandoned for its salvage value S_5.

$$V_5(\tau + T_3) \;=\; \sup_{\tau+T_3 \le \tau_A \le \tau+T_4} \tilde{\mathbb{E}}\left[e^{-r(\tau_A-\tau-T_3)} \max\{V_5(\tau_A),\, S_5(\tau_A)\}\right],$$

$$V_5(\tau + T_4) \;=\; \max\{V_9(\tau + T_4),\, V_8(\tau + T_4) - I_4,\, S_5(\tau + T_4)\}, \qquad (4.74)$$

$$S_5(\tau_A) \;=\; 0.5 \cdot \left[I_1 e^{-d(\tau_A)} + I_2 e^{-d(\tau_A-T_2)} + (I_3 - \tilde{I}_3)e^{-d(\tau_A-T_3)}\right].$$

Switch 6 → 10 at time τₐ:
After having reached states 6, 7, 8, or 9 the only flexibility left up to the planning horizon $\tau + T_5$ is the option to abandon. Therefore the only thing left to determine is the salvage value for each of the states 6 to 9. It is important to note that the project values V_i for $i = 6, \ldots, 9$ depend on the static NPV $V(t)$ of the project alone. The project values reflect the previously taken decisions to contract, operate full, or expand the scale which one might have expected already from the project values V_4 and V_5.

$$V_6(\tau + T_4) \qquad\qquad\qquad\qquad\qquad\qquad\qquad\qquad\qquad (4.75)$$
$$= \sup_{\tau+T_4 \le \tau_A \le \tau+T_5} \tilde{\mathbb{E}}\left[e^{-r(\tau_A-\tau-T_4)} \max\{(1+e)V(\tau_A),\, S_6(\tau_A)\}\right]$$
$$S_6(\tau_A)$$
$$= 0.5 \cdot \left[I_1 e^{-d(\tau_A)} + I_2 e^{-d(\tau_A-T_2)} + I_3 e^{-d(\tau_A-T_3)} + I_4 e^{-d(\tau_A-T_4)}\right].$$

Switch 7 → 10 at time τₐ:

$$V_7(\tau + T_4) \qquad\qquad\qquad\qquad\qquad\qquad\qquad\qquad\qquad (4.76)$$
$$= \sup_{\tau+T_4 \le \tau_A \le \tau+T_5} \tilde{\mathbb{E}}\left[e^{-r(\tau_A-\tau-T_4)} \max\{V(\tau_A),\, S_7(\tau_A)\}\right],$$
$$S_7(\tau_A)$$
$$= 0.5 \cdot \left[I_1 e^{-d(\tau_A)} + I_2 e^{-d(\tau_A-T_2)} + I_3 e^{-d(\tau_A-T_3)}\right].$$

Switch 8 → 10 at time τₐ:

$$V_8(\tau + T_4) \qquad\qquad\qquad\qquad\qquad\qquad\qquad\qquad\qquad (4.77)$$
$$= \sup_{\tau+T_4 \le \tau_A \le \tau+T_5} \tilde{\mathbb{E}}\left[e^{-r(\tau_A-\tau-T_4)} \max\{(1-c)(1-e)V(\tau_A),\, S_8(\tau_A)\}\right],$$
$$S_8(\tau_A)$$
$$= 0.5 \cdot \left[I_1 e^{-d(\tau_A)} + I_2 e^{-d(\tau_A-T_2)} + (I_3 - \tilde{I}_3)e^{-d(\tau_A-T_3)} + I_4 e^{-d(\tau_A-T_4)}\right].$$

Switch $9 \rightarrow 10$ *at time* τ_A:

$$V_9(\tau + T_4) \tag{4.78}$$

$$= \sup_{\tau + T_4 \leq \tau_A \leq \tau + T_5} \tilde{\mathbb{E}} \left[e^{-r(\tau_A - \tau - T_4)} \max\{(1-c)V(\tau_A), \ S_9(\tau_A)\} \right]$$

$$S_9(\tau_A) = $$

$$0.5 \cdot \left[I_1 e^{-d(\tau_A)} + I_2 e^{-d(\tau_A - T_2)} + (I_3 - \tilde{I}_3) e^{-d(\tau_A - T_3)} \right].$$

Turning now to the well posedness of the optimization model, it is easy to see that none of the value functions V_i of the different states will blow up due to the finite planning horizon. Furthermore, since no explicit cash flow functions needed to be taken into account and the GBM of the static NPV poses no difficulties, all optimization problems are given in a way that allows application of the verification theorem of the generalized stopping model above, which holds for combinations of stopping models as well.

The next step is to translate the formulated structure of the investment project into a graphical representation using the elements introduced above. While there are no states in Trigeorgis' model that are costly reversible, all the incorporated states have to be represented as single states. By noting that the states 4 and 5 as well as 6 to 9 are parallel states, they can be grouped in order to simplify the decomposition graph. Furthermore, there are two kinds of switches to the abandonment state which need to be distinguished. The first denotes abandoning without recovery of the investment outlay and the second denotes the American option to abandon for salvage value during the operating stage indexed by A. The reason for not modelling the abandoned state 10 in the same way as the others is that the abandonment option is inherent in almost any investment project and can be exercised during the whole life of the project. The resulting graphical decomposition is displayed in Figure 4.14.

The model is now sufficiently specified to implement a numerical solution technique to find the value of the project including all involved real options. Numerical results derived by using a binomial (trinomial) tree can be found in Trigerogis [225]. While the stochastic control framework considered here takes a continuous time approach, we will concentrate on numerical solution techniques on the level of the corresponding system of PDEs rather than employing a discrete time approach like binomial trees. A detailed example will be given in Chapter 6.

Although the model of Trigeorgis already contains several interacting real options its decomposition graph exhibits a quite simple structure. The graphical decomposition method admits an intuitive and clear representation of option interactions. As it was shown in the previous examples it is easy to implement and provides a structured procedure which gives further insights into the valuation process. Furthermore, the graphical representation can be easily analyzed and adjusted by adding or removing real options. Because the decision problem represented in the graph is already structured in a way that allows for easy numerical implementation, it may serve as a graphical modelling tool for numerical real option valuation packages.

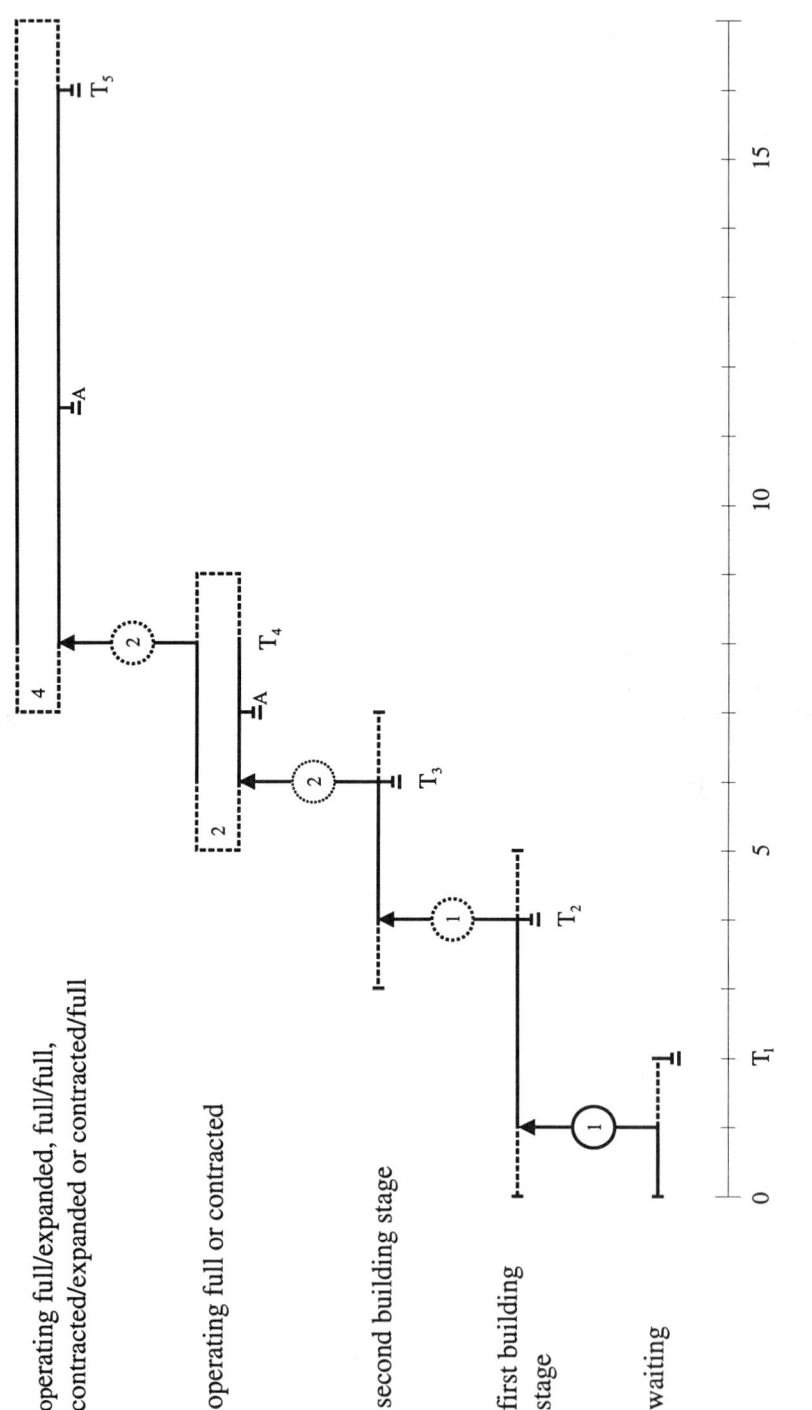

Figure 4.14: Option interactions

Most important, the proposed graphical decomposition method is very user friendly because it does not require a deeper understanding of the complex mathematics behind the graphical representation. A rough idea about the essence of option pricing should be sufficient to frame different applications that incorporate complex real option interactions. Moreover, the graphical approach is easy to communicate to decision makers in that it can be understood quickly and focusses on the main points rather than stressing technicalities. It is therefore an important contribution to the real option valuation process discussed in Chapter 3.

We have seen that the contingency structure of many real option models mentioned in the literature can be expressed in terms of the graphical decomposition method. Since it was not discussed how to find either analytical or numerical solutions for these models, this will be subject to the following chapters. Moreover, some additional real options' characteristics need to be embedded in the graphical decomposition model, e.g., how to represent competition, time lags, learning effects, forced switches by random events etc.

4.3 Example: Timing and Intensity of Investment

Closed form solutions for value functions when real options are involved are rare. As for most financial options the usual numerical techniques have to be employed. That is one of the reasons why classical real option research mostly considers isolated options of either European or perpetual American type. For these option types, closed form solutions can be sometimes found. In a similar fashion the option to choose the intensity of investment is now presented as a perpetual American option in order to analyze its properties. As it will turn out, the option to choose the intensity of investment may add tremendous value to the project in comparison to the isolated timing option where the intensity of investment is a priori fixed and there is no flexibility with respect to the intensity. However, the intensity option is often neglected in real option pricing, although it seems that in most real-world cases where flexibility with respect to timing is present, flexibility with respect to intensity is naturally present as well — at least to some degree. It is surprising that the problem of timing and intensity was addressed just recently by Bar-Ilan/Strange [15]. Although their model focusses on microeconomic issues, we adopt and adjust it to the proposed stochastic control model here. The problem of timing and intensity will prove very useful to demonstrate how to deal with the stochastic control framework and how one can find closed form solutions for isolated real options of perpetual American type. Beyond that it will highlight the importance of the intensity option. The example is structured in the following way. First, we will work through steps 1 to 6 of the graphical decomposition method. Afterwards the solution of the stochastic control problem will be derived. The solution is analyzed in comparison to the pure timing option and a sensitivity analysis is performed.

Consider a firm that has the opportunity to invest in a new plant. The plant produces a single product whose price P fluctuates stochastically subject to the geometric Brownian motion

$$\mathrm{d}P_t = (r - \delta)\, P_t\, \mathrm{d}t + \sigma\, P_t\, \mathrm{d}\tilde{B}_t, \qquad P_0 = p \qquad (4.79)$$

under the equivalent martingale measure $\tilde{\mathbb{P}}$ where r, δ and σ are assumed to be constant over time and are defined as above. For simplicity the firm is assumed to be a profit maximizing price taker. Building the plant with theoretical marginal capacity M requires an initial investment outlay per unit capacity of m. The production rate function of the plant is M^a for $0 \leq a \leq 1$, which means that the output efficiency of the plant is less than or equal to 1 and is a concave function of marginal capacity M.[28] The instantaneous variable cost for one unit of output is c. Building the plant that produces one unit per time requires an initial investment outlay of m. After having invested in the plant the firm produces at a constant rate of M^a forever.

The question that is to be answered now is how this simultaneous timing and intensity option can be represented in terms of the graphical decomposition method and how the optimal value function for this option can be derived.

Working through steps 1 to 6 as above, we start by defining two states. The firm is either inactive — state 0 — or operating with intensity M — state 1. Since M can be chosen continuously the number of states is strictly speaking infinite. However, it suffices to reduce the situation to only the two mentioned states. For both states the first possible entry time is zero. The latest possible exit time for state 0 is infinity since the investment option may never be exercised. On the other hand there exists no exit time from state 1 because the firm operates forever after having decided to invest. The underlying uncertainty for both states is the output price P_t with given starting value $P_0 = p$ at time 0. By assuming that P_t is either the price of a traded commodity or that spanning holds the equivalent martingale measure, $\tilde{\mathbb{P}}$ can be determined. The instantaneous cash flow f_0 earned while the firm is inactive is zero. If the firm is operating at M it earns an instantaneous cash flow of

$$f_1(t, P_t, M) = f_1(P_t, M) = M^a(P_t - c) \qquad (4.80)$$

which does not depend on t explicitly since the perpetual problem is time homogeneous. As for all options to invest there is only one irreversible switch to consider which is of American type. Furthermore, the option is perpetual because it can be exercised at any time now or in the future. Irreversible switching is maximized over all possible values of M when exercising is triggered. Therefore, the action space $\mathcal{Z}$ is the interval $(0, \infty)$ and $M \in \mathcal{Z}$. The switching cost function H^τ is given by

$$H^\tau(t, P_t, 0, M) = H^\tau(M) = m\, M. \qquad (4.81)$$

---

[28]This inefficiency of the production process may have several reasons. First, it may not be optimal from a production cost view to operate at full scale but somewhat below. Second, physical reasons may limit the output efficiency of the production process, e.g., the generation of scrap.

In terms of the graphical decomposition method the combined timing and intensity option can be displayed as in Figure 4.15. As above the closed circle represents

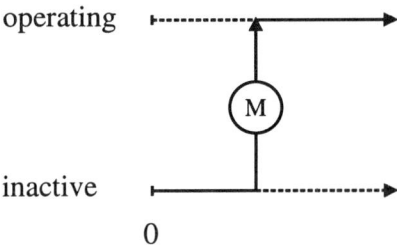

Figure 4.15: Perpetual timing and intensity option

the American type option. However, M in the circle is now a continuous variable rather than a certain number of states over which maximization takes place.

Turning now to the solution and analysis of the generalized timing option we first state the optimization problem for which a solution has to be found. In order to stress the importance of the opportunity to simultaneously choose the timing and intensity of investment we will discuss the pure timing option for fixed M in comparison as well.

While the firm starts in state 0 the optimization problem of the generalized timing option according to Section 3.4 is

$$V_0(P_0) = \sup_{(\tau,M)\in[0,\infty)\times\mathcal{Z}} \tilde{\mathbb{E}}\left[\int_\tau^\infty e^{-rt} M^a (P_t - c)\, dt - e^{-r\tau} mM\right]$$

$$= \sup_{(\tau,M)} \tilde{\mathbb{E}}\left[e^{-r\tau} V_1(P_\tau, M) - e^{-r\tau} mM\right]. \tag{4.82}$$

If the firm is already operating there is no additional flexibility to alter the state of the firm. Therefore, the value of the operating firm $V_1(P, M)$ can be calculated as

$$V_1(P, M) = \tilde{\mathbb{E}}\left[\int_0^\infty e^{-rt} M^a (P_t - c)\, dt\right] \tag{4.83}$$

$$= \int_0^\infty e^{-rt} M^a (P e^{(r-\delta)t} - c)\, dt$$

$$= M^a \left(\frac{P}{\delta} - \frac{c}{r}\right).$$

In order to determine the value of the inactive firm $V_0(P)$ we use Itô's lemma to get the fundamental pricing equation in the continuation region C^τ. For the time homogeneous problem considered here, V_0 has to satisfy the ordinary differential equation (ODE)

$$\frac{1}{2}\sigma^2 P^2 \frac{d^2 V_0}{dP^2} + (r - \delta) P \frac{dV_0}{dP} - rV_0 = 0. \tag{4.84}$$

The general solution to this ODE is given by

$$V_0(P) = A P^{\beta_1} + B P^{\beta_2} \tag{4.85}$$

where A and B are constants to be determined. β_1 and β_2 can be obtained by substituting the general solution (4.85) into the ODE (4.84). The resulting characteristic equation

$$\frac{1}{2}\sigma^2\beta(\beta - 1) + (r - \delta)\beta - r = 0 \tag{4.86}$$

has the roots

$$\beta_{1/2} = \frac{1}{2} - \frac{r - \delta}{\sigma^2} \pm \sqrt{\left(\frac{r - \delta}{\sigma^2} - \frac{1}{2}\right)^2 + \frac{2r}{\sigma^2}} \tag{4.87}$$

where we have $\beta_1 \geq 1$ and $\beta_2 \leq 0$. Furthermore, define the maximum operator when switching is triggered as

$$\mathcal{M}^\tau V_0(P) = \max_M \{V_1(P, M) - mM\} \tag{4.88}$$

which is independent of t for the time homogeneous problem. Therefore, the continuation region $\mathcal{C}^\tau$ takes the form

$$\mathcal{C}^\tau = \{P \in \mathbb{R}_+;\ V_0(P) > \mathcal{M}^\tau V_0(P)\}. \tag{4.89}$$

The interpretation is quite simple. Whenever the process P_t reaches a yet to be determined trigger price $P^*(M^*)$, the firm starts operating with an optimal capacity M^*. In order to solve for V_0 first note that as P_t approaches zero the option to invest becomes worthless, i.e.,

$$\lim_{P \to 0} V_0(P) = 0. \tag{4.90}$$

Since β_2 is negative we must have $B = 0$ for V_0 to go to zero as P decreases. Furthermore, we require V_0 to be stochastically C^2 which results in the usual value matching and smooth pasting conditions for optimal stopping at the trigger price P^* where M is kept fixed for a while.

$$V_0(P^*(M)) = V_1(P^*(M), M) - mM, \tag{4.91}$$
$$V_0'(P^*(M)) = V_1'(P^*(M), M). \tag{4.92}$$

Substituting (4.85) with $B = 0$ into (4.91) and (4.92) yields two equations which can be solved for P^* and A.

$$A P^{*\beta_1} = M^a \left(\frac{P^*}{\delta} - \frac{c}{r}\right) - mM, \tag{4.93}$$

$$A\beta_1 P^{*\beta_1 - 1} = \frac{M^a}{\delta}. \tag{4.94}$$

Thus, we find that for fixed M,

$$P^*(M) = \frac{\beta_1}{\beta_1 - 1} \frac{\delta}{r} \left(c + rmM^{1-a} \right), \tag{4.95}$$

$$V_0(P) = \frac{M^a(c + rmM^{1-a})}{r(\beta_1 - 1)} \left(\frac{P}{P^*(M)} \right)^{\beta_1}. \tag{4.96}$$

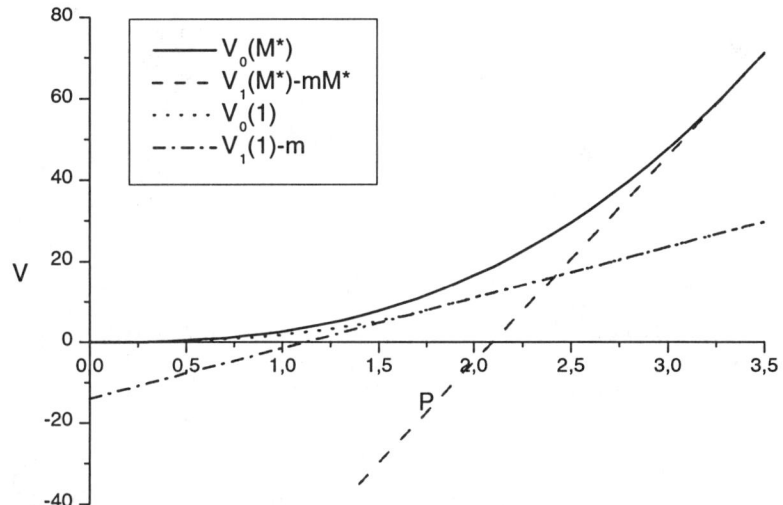

Figure 4.16: Comparison of option values for fixed and variable intensity with base case parameters $\sigma = 0.15$, $r = 0.1$, $\delta = 0.08$, $c = 1.0$, $m = 4.0$ and $a = 0.5$. Fixed intensity (pure timing option) with $M = 1$ and corresponding optimal trigger price $P^* = 1.812$. Timing and intensity with optimal capacity $M^* = 16.38$ and corresponding trigger price $P^*(16.38) = 3.390$.

Next the maximum operator is employed to find the optimal capacity level M^* which maximizes $V_0(P^*(M))$. The solution is

$$M^{*1-a} = \frac{a}{\beta_1(1-a) - 1} \frac{c}{rm} \tag{4.97}$$

as long as $\beta_1(1-a) > 1$. Otherwise an optimal capacity level $M^* > 0$ does not exist and the investment option will never be exercised. The corresponding trigger price $P^* = P^*(M^*)$ is given by

$$P^*(M^*) = \frac{\beta_1(1-a)}{\beta_1(1-a) - 1} \frac{\delta}{r} c. \tag{4.98}$$

It is now easy to show that the derived value function V_0 satisfies the verification theorem of Section 3.4.2 and that V_0 is indeed the optimal value function of

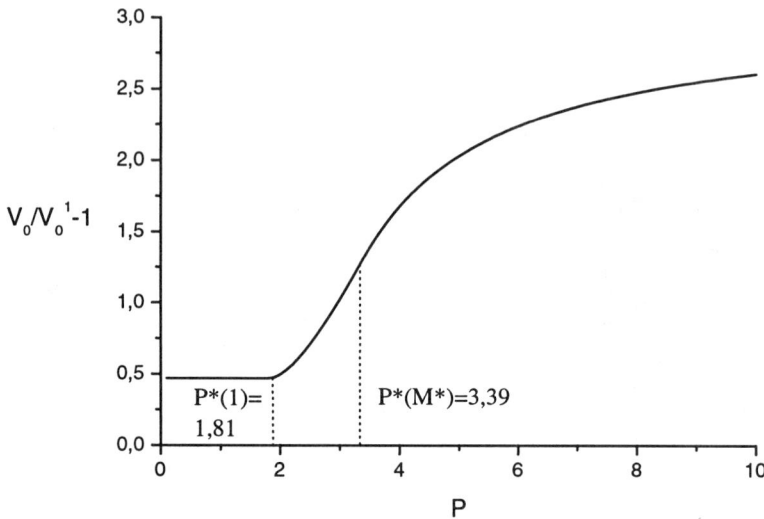

Figure 4.17: Percentage of added value by intensity option with optimal capacity $M^* = 16.38$ and corresponding trigger price $P^*(16.38) = 3.390$ in comparison to the pure timing option with $P^*(1) = 1.81$.

the above generalized stopping problem. V_0 is by construction stochastically C^2. Moreover, it satisfies the following system of quasi-variational inequalities:

$$\frac{1}{2}\sigma^2 P^2 \frac{d^2 V_0}{dP^2} + (r - \delta)P\frac{dV_0}{dP} - rV_0 \leq 0$$

$$V_0(P) \geq \mathcal{M}^\tau V_0(P). \qquad (4.99)$$

One of the inequalities is an equality.

In C^τ the first inequality is in fact an equality and outside C^τ the second is an equality. From that it follows directly that V_0 is optimal according to the verification theorem of Section 3.4.2.

In the sequel the results when timing and intensity are considered simultaneously are compared to those of the pure timing option with fixed capacity level as discussed in Dixit/Pindyck [72]. For the pure timing option we fix the capacity level to $M = 1$, i.e., the firm has only the flexibility to invest in a plant with unit capacity. The corresponding trigger price $P^*(1)$ and the value function $V_0^1(P)$ can be determined by equations (4.95) and (4.96) for $M = 1$,

$$P^*(1) = \frac{\beta_1}{\beta_1 - 1}\frac{\delta}{r}(c + rm), \qquad (4.100)$$

$$V_0^1(P) = \frac{c + rm}{r(\beta_1 - 1)}\left(\frac{P}{P^*(1)}\right)^{\beta_1}. \qquad (4.101)$$

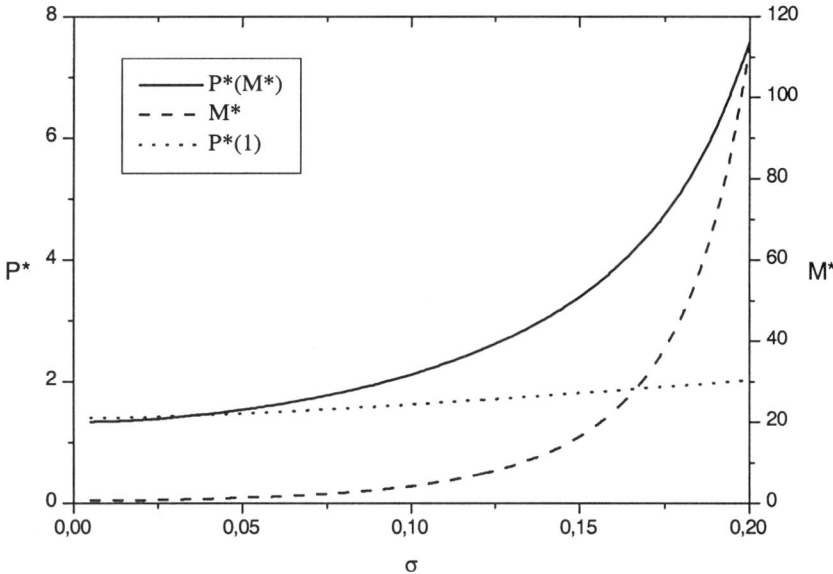

Figure 4.18: Impact of changing price volatility σ on trigger price $P^*(M^*)$, $P^*(1)$ and optimal capacity M^* for base case parameters.

In order to isolate the effect of the control opportunity to choose the intensity of investment in comparison to the pure timing option, a numerical analysis of the option values is performed. For the base case parameters $\sigma = 0.15$, $r = 0.1$, $\delta = 0.08$, $c = 1.0$, $m = 4.0$ and $a = 0.5$ the option values for both models are displayed in Figure 4.16. As can be seen from the figure the result is striking. First of all, when intensity of investment is allowed for, the option value is significantly higher than for the pure timing option. It is obvious that the neglect of the intensity option that usually occurs in the literature leads to a tremendous misvaluation. Since intensity options often appear as naturally as timing opportunities in real-world decision problems, there can be enormous value gained by correctly considering intensity in addition to timing of investment. For the base case parameters the percentage value added by taking intensity into account is presented in Figure 4.17. Although this might be an extreme example due to the large difference between $M^* = 16.38$ and $M = 1$, the relationship holds true in general.

In the interval $(0, P^*(1)]$ before the pure timing option is exercised the value added by the intensity option is constant by 46.9% of the pure timing option. In the interval where exercise of the pure timing option has already been triggered,

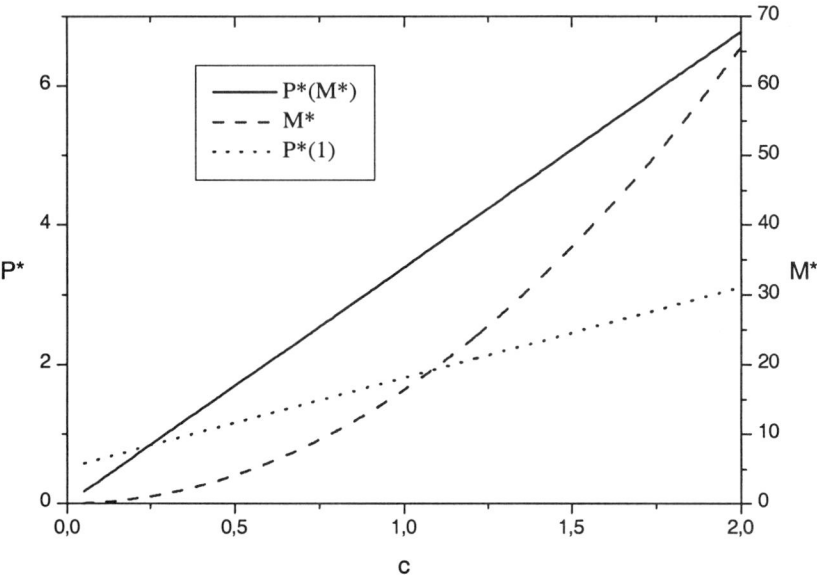

Figure 4.19: Impact of changing marginal production cost c on trigger price $P^*(M^*)$, $P^*(1)$ and optimal capacity M^* for base case parameters.

this ratio is convex and increases further up to the turning point at $P^*(M^*)$ after which the function is concave and still increasing. In extreme for large prices the value gained by considering intensity is more than 250%!

Second, the trigger price $P^*(M^*)$ is in general higher than the trigger price $P^*(1)$ of the pure timing option. Furthermore, M^* is much larger than 1. This seems to be counterintuitive since with the additional degree of flexibility to choose the intensity, one would have expected that with a rising trigger price P^* the installed optimal capacity M^* should decrease. However, the opposite is true. This phenomenon will become clear by a discussion of the comparative statics of the trigger price and the optimal installed capacity. For ease of exposition we will concentrate on the numerical example rather than analyzing the general formulas which can, however, easily be derived.

Starting with the impact of the volatility σ on the trigger price an increase in volatility results as expected in an increase of P^* for both models. The effect of the volatility σ on M^*, however, is surprising. An increase in uncertainty leads to more investment when it finally occurs. An explanation of this result can be

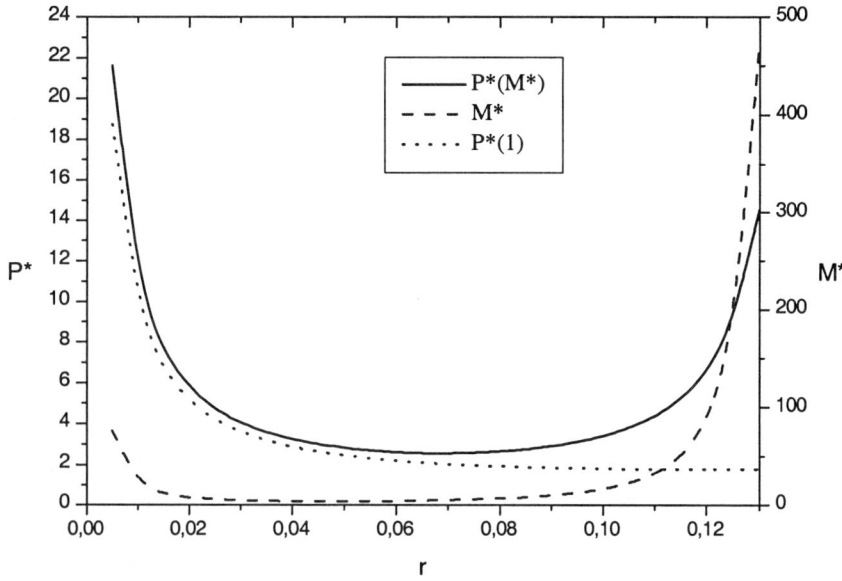

Figure 4.20: Impact of changing interest rate r on trigger price $P^*(M^*)$, $P^*(1)$ and optimal capacity M^* for base case parameters.

found by separating the effect of timing and intensity. The timing is determined by the value matching and smooth pasting condition (4.91) and (4.92). At P^*, the marginal cost and the marginal benefit of a short delay are equal. The optimal intensity of investment is determined by maximizing $V_0(P)$ with respect to M. At M^*, the marginal cost and the marginal benefit of additional capacity are equal. An increase of uncertainty raises the option value of delay and consequently raises P^* for both models as can be seen from Figure 4.18. On the other hand when P^* is higher, the marginal benefit of a further unit of capacity rises, leading to an increase of M^*. Thus, higher uncertainty has two effects. The first is discouraging investment by a higher trigger price, and the second is encouraging investment by higher optimal intensity.

A similar observation can be made for the marginal production cost c shown in Figure 4.19. The trigger price of both models rises linearly in c, see equations (4.98) and (4.100). As before the delay in investment caused by the higher value of waiting leads to higher intensity of investment when it occurs.

The comparative statics behind the other parameters exhibit different effects,

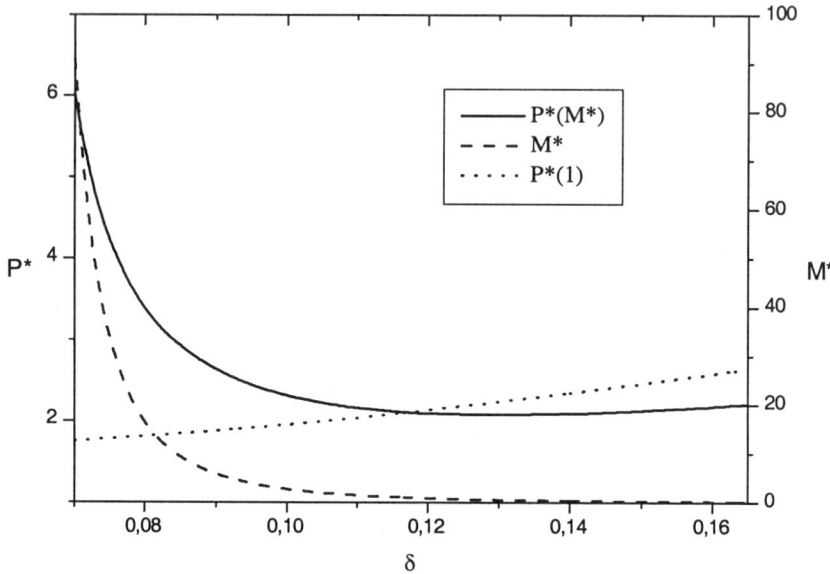

Figure 4.21: Impact of changing rate of return shortfall δ on trigger price $P^*(M^*)$, $P^*(1)$ and optimal capacity M^* for base case parameters.

highlighting also the differences between the pure timing option and the timing and intensity option. The riskless interest rate r for example leads as expected to a decrease in the trigger price of the pure timing option. The impact of the interest rate on the timing and intensity option however is ambiguous, see Figure 4.20. As long as r is small both the trigger price as well as the optimal capacity decrease. This is due to the reduction of the marginal cost of production c/r for rising r. However, this effect is offset by the increasing cost of installed capacity $r m M^*$ leading first to rising optimal capacity levels M^* and subsequently to a rising trigger price $P^*(M^*)$.

One would expect that the opposite is true for the rate of return shortfall δ. But this is only valid for the pure timing option. An increase in δ leads to higher investment triggers $P^*(1)$. In contrast, the trigger price for the timing and intensity option first goes down before it reaches a minimum and rises again because of the nonlinear dependence of β_1 on δ together with the output efficiency a. The impact of δ on M^* is more intuitive. The optimal capacity level declines with increasing rate of return shortfall due to the monotonicity of β_1 in δ. The result is summarized in Figure 4.21.

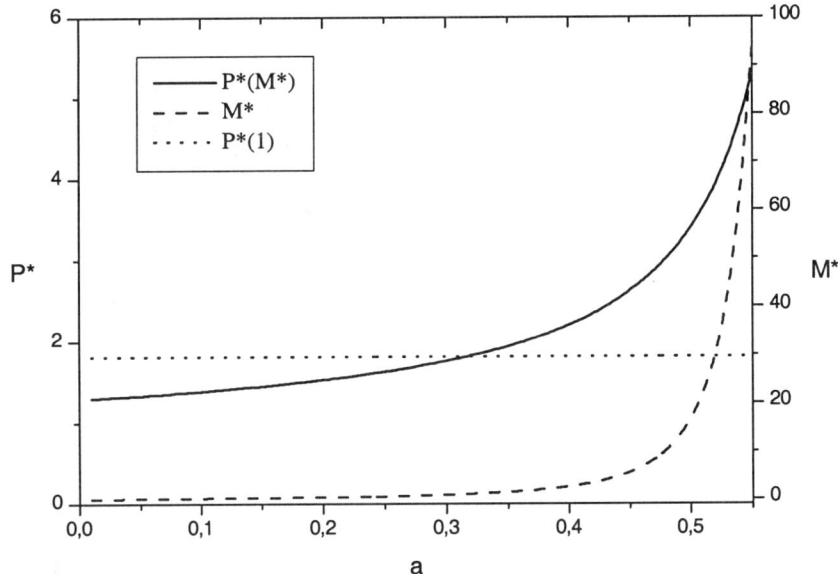

Figure 4.22: Impact of changing output efficiency a on trigger price $P^*(M^*)$, $P^*(1)$ and optimal capacity M^* for base case parameters.

As a lower rate of return shortfall, a rising output efficiency a directly increases the marginal benefit of capacity relative to its marginal cost, which results in higher optimal capacity levels. Thus, higher efficiency encourages investment as expected. The co-movement of P^* and M^* is observable here as well. The cost of waiting has to be weighed against the higher irreversible commitment for more capacity. This discourages investment and leads to a higher trigger price.

Finally, the unit cost of capacity m affects the trigger price of the pure timing option in a linear way. By way of contrast, a rise in m is completely compensated for by a corresponding decrease in the optimal capacity level M^* for this particular model of timing and intensity. As a consequence the trigger price is invariant with respect to m. Of course, this might not be the case for other output efficiency functions.

To conclude, we have seen that the sole concentration on timing opportunities in almost all real option models considered in the literature can lead to tremendous misvaluation of option values. The simultaneous optimization of investment strategies with respect to both timing and intensity provides a vital extension to the real

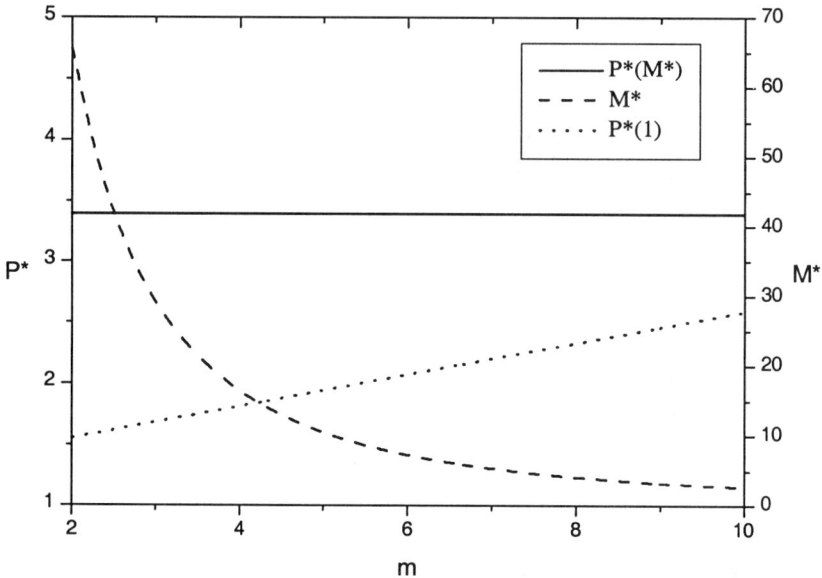

Figure 4.23: Impact of changing unit cost of capacity m on trigger price $P^*(M^*)$, $P^*(1)$ and optimal capacity M^* for base case parameters.

options literature. The results produced by the timing and intensity option seemed to be counterintuitive. The flexibility to choose the intensity of investment raises both the investment trigger price and the optimal capacity. Therefore, this strategy delays investment even further. It was shown that the timing and intensity option can be easily embedded as well into the stochastic control approach to real option pricing as into the graphical decomposition method. The timing and intensity option was used as an example to demonstrate how the stochastic control framework can be used to find analytic solutions for the value function of a generalized timing option of perpetual type.

Chapter 5

Extensions: Competition and Time Delay Effects

So far the discussion has dealt with several examples of how to use the stochastic control framework to value real options, and how to display real option interactions by means of the graphical representation of the contingency structure. In this chapter several recent advances in real option pricing are included in the framework. Paralleling the conceptual framework of real options of Section 2.2.2, some assumptions will be relaxed in order to take account of competition and time delay effects in the stochastic control framework. The presentation of these topics will concentrate on the application of the methods rather than on mathematical thoroughness. Thus, there are no proofs provided for the extensions of the generalized timing and switching options of Sections 3.3 and 3.4. Although some of the necessary proofs would not be trivial, their contribution to the valuation and analysis of option interactions we focus on would be limited.

5.1 Competitive Interaction

As already mentioned in Section 2.2.2, many real options that are embedded in an investment are not exclusive to only one firm but may also be shared by other competitors. This would not cause too much difficulties if these option were separable from the investment and could be sold solely in liquid markets. Unfortunately, most real options are tied to the corresponding project and cannot be sold without selling the investment project itself. Consider for example the option to abandon a project in midstream. The option is naturally tied to the project. There is no means to sell it to another firm or investor because, separated from the project, it is worthless. On the other hand the shared option to invest in a new geographic market for instance would not be purchased by other firms unless there exist some severe market entry barriers or contractual rights that only enable the holding firm to enter the market. But in our interpretation this would be an exclusive option that is not exposed to competitive erosion. Put together, competition together with non-tradability of shared options might reduce or even eliminate the value of real options. The question to be answered is what consequences does this have for valuing real options.

To start with we distinguish three dimensions of competition. The first is the market structure. Depending on the number of competitors, the market structure can be either a monopoly, an oligopoly or perfect competition if many market participants are present. The second dimension depends on the subject of investigation. A decision maker can be interested either in the optimal decision of the single firm or in the outcome of the decisions of all market participants. The latter results in microeconomic equilibrium theory which will not be subject to further discussion.[1] The former gives rise to the question how to value shared real options in a competitive environment and how to find the optimal decision strategies for an individual firm. Depending on the market structure, different ways to model competition can be distinguished, as already discussed in 2.2.2. Competitive effects can be regarded as either deterministic or stochastic. Furthermore, depending on the market structure competitive actions may be exogenous or endogenous to the firm. In sufficiently perfect markets it is reasonable to assume that actions by the firm do not cause direct strategic answers by other firms. Therefore, competition can be modelled as exogenous, possibly random, events. On the other hand in oligopolistic markets actions taken by the firm may likely result in strategic answers by its competitors. For that reason the firm has to take account of possible reactions of competitors when choosing its strategy. In this case competition can be regarded as endogenous, usually resulting in a combination of real option analysis and game theory.

For both concepts, exogenous and endogenous competition, it will be shown below how they can be modelled in the stochastic control framework and how they can be represented in the graphical decomposition method.

5.1.1 Exogenous Competition

In case competition is exogenously given, the firm has to weigh the value of waiting against the possible erosion of value by competitors' actions which it cannot influence. The firm has to determine what information about competition is available to the it. If for example the firm knows in advance the strategies of its competitors and their impact on the firm's value function, the situation is completely deterministic. In this quite unrealistic case the firm may react by preempting competitors' actions and putting up with the loss in value by early exercise of its option.[2] On the other hand the case where other firms randomly enter a market and exercise their real

---

[1] See the extensive literature on equilibrium theory, e.g., chapters 8 + 9 of Dixit/Pindyck [72] and the references therein.

[2] See Trigeorgis [221, 225] for a discussion of preemption strategies.

option to invest seems to be more realistic. The firm might have a rough idea about the intensity of competition and its impact without having full information about when and how other firms act.

A stylized example that will be discussed in what follows is the option to invest in a new market that is shared among many competitors. The firm has to decide when to exercise its option to wait and enter the market by taking account for the possible loss of early mover advantages by being late in the market. For simplicity it is assumed that with each competitor entering the market the project value of the firm is reduced by a certain percentage g. Therefore the impact of competition is known.[3] What is random to the firm is the time when its competitors choose to exercise their investment option. Such a situation is conveniently modelled by a Poisson process which takes account of the random points in time when the gross project value V that may be earned by the firm when entering the market drops.[4] The question that arises now is what optimal strategy the firm should pursue when it faces random competitive entry. Different from what is usually argued in the literature,[5] we find that a preemption strategy is not always feasible in case of exogenous stochastic competition. On the contrary, it very much depends on whether the additional risk introduced by competition is priced at capital markets or if it is private to the firm. In order to highlight the importance of correct modelling of the relevant sources of uncertainty, we will compare these two cases numerically.

Before starting with the mathematical formulation of the decision situation the corresponding graphical representation for the problem is developed. Assume that a company possesses the shared option to invest in a new market with gross market value V following geometric Brownian motion. Furthermore, the firm faces exogenous uncertain competition where the arrival rate of an additional competitor is given by the parameter λ. Whenever a new competitor enters the market a drop in gross market value that is available to the firm to $(1 - g)V_t$ is induced. The next competitor to enter causes another drop to $(1 - g)^2 V_t$ and so forth. Therefore the gross market value available to the firm follows the mixed jump diffusion process

$$\mathrm{d}V_t = \mu \, V_t \, \mathrm{d}t + \sigma \, V_t \, \mathrm{d}B_t - g \, V_t \mathrm{d}q_t, \qquad V_0 = v \qquad (5.1)$$

under the probability measure $\mathbb{P}$ in the original world. The problem of how to find the equivalent martingale measure for this process will be discussed below. The parameters μ and σ are assumed to be constant and positive. The parameter g of course is in the interval $[0, 1]$. The Poisson process $\mathrm{d}q_t$ is defined as

$$\mathrm{d}q_t = \begin{cases} 1 & \text{with probability} \quad \lambda \, \mathrm{d}t, \\ 0 & \text{with probability} \quad 1 - \lambda \, \mathrm{d}t, \end{cases} \qquad (5.2)$$

---

[3]There exists no principle obstacle to relax this assumption and introduce further information asymmetry by assuming for example a probability distribution over the impact of competitors' actions when they occur, see e.g., Lambrecht /Perraudin [135].

[4]See Section 3.2.2 for the introduction of mixed Poisson jump and diffusion processes.

[5]See Trigeorgis [221], p. 154.

with λ constant and positive. If the firm wants to enter the market investment costs of I are incurred. The investment opportunity is assumed to be perpetual.

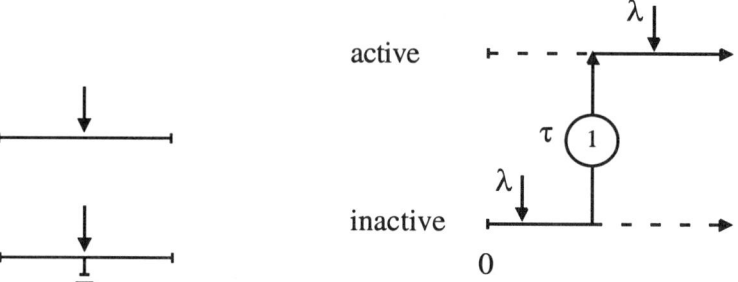

Figure 5.1: Modelling element exogenous competition. Arrow represents exogenous competitive erosion. Lower graph displays the case where competition makes the option worthless.

Figure 5.2: Graphical representation of the option to invest under competition. λ indicates Poisson process.

Similar to the timing and intensity option there are two states to identify — active and inactive — between which one irreversible switch can be performed. For both states the first possible entry time is zero and the latest possible exit time to state inactive is infinity. The underlying source of uncertainty for both states is the readily available part of the gross market value V_t. There is no explicit instantaneous cash flow f_i modelled. However cash payments to the active firm are taken account of by the dividend yield δ as in Section 4.2. Irreversible switching from inactive to active takes place at a stopping time τ where the switching costs $H^\tau = I$ are incurred. In order to represent this shared option to invest in the graphical decomposition, we introduce the modelling element *exogenous competition* in Figure 5.1. The arrow means competitive erosion. The corresponding graph for the considered option to invest can be found in Figure 5.2. The Poisson process is indicated by the mean arrival rate λ. Furthermore, competitive action is present while the firm is either inactive or active. Thus, the arrow standing for competition is displayed for both states.

What remains to investigate is how the additional risk that comes along with the Poisson jumps should be treated from a capital market point of view. Without going too far into detail there are two cases thinkable. First, the Poisson risk is systematic.[6] This means that capital markets assign a risk premium to the Poisson risk because it cannot be diversified away by holding other assets. In order to determine the equivalent martingale measure $\tilde{\mathbb{P}}^1$ for this case we proceed in the same way as in Section 4.2. Suppose there exists a twin asset M that completely duplicates the risk of V. The only difference is in the cash flow stream (or rate of

---

[6]As assumed in Trigeorgis [221] and [225], p. 285.

return shortfall) δ earned by the operating project. Then M follows the stochastic process

$$dM_t = (\mu + \delta) M_t \, dt + \sigma M_t \, dB_t - g \, dq_t. \tag{5.3}$$

Furthermore, under the equivalent martingale measure the discounted asset price process needs to be a martingale. Alternatively, the discounted project value earns a below equilibrium rate of return shortfall δ. In order to find the associated asset price for risk that determines the new Brownian motion, we equate the discounted project value price process

$$
\begin{aligned}
d(e^{-rt} V_t) &= e^{-rt}[-rV_t \, dt + dV_t] \\
&= e^{-rt} [(\mu - r)V_t \, dt + \sigma V_t \, dB_t - gV \, dq_t] \\
&= e^{-rt} V_t [(\mu - r) \, dt + \sigma \, dB_t - dq_t] \\
&= e^{-rt} V_t [(\mu - r - \lambda g) \, dt + \sigma \, dB_t] \\
&= e^{-rt} V_t \sigma \left[\frac{\mu - r - \lambda g}{\sigma} \, dt + dB_t \right]
\end{aligned}
\tag{5.4}
$$

where it was taken account for dq_t to be 1 with probability λdt leading to a drop of $100g\%$ in project value V. The market price of risk including the Poisson risk component is then given by

$$\vartheta_t = \frac{\mu - r - \lambda g}{\sigma} \tag{5.5}$$

determining the new Brownian motion $\tilde{B}_t$ under the equivalent martingale measure $\tilde{\mathbb{P}}^1$

$$d\tilde{B}_t = \vartheta_t \, dt + dB_t. \tag{5.6}$$

Together with the condition

$$\tilde{\mathbb{E}}^1 \left[d(e^{-rt} V_t) \right] = -e^{-rt} \delta V \, dt \tag{5.7}$$

this yields the project value dynamics under the equivalent martingale measure $\tilde{\mathbb{P}}^1$,

$$dV_t = (r + \lambda g - \delta) V_t \, dt + \sigma V_t \, d\tilde{B}_t - g \, dq_t. \tag{5.8}$$

The interpretation is quite simple. In the risk neutral world all investors require an expected rate of return on an asset equal to the risk free rate of return r minus intermediate cash flows δ. The additional term λg takes care of the expected drop in value by the Poisson component.

The second case to be considered is the situation where the Poisson risk is assumed to be private to the firm, i.e., it can be diversified by other investors for example by investing in direct competitors. Therefore investors will not demand an extra risk premium for the risk associated with the Poisson component. However, the uncertainty about the future gross market value represented by the volatility σ still demands a risk premium. So the twin asset replicating the uncertainty with respect to the gross market value has the asset dynamics

$$dM_t = (\mu + \delta)\, M_t\, dt + \sigma\, M_t\, dB_t. \tag{5.9}$$

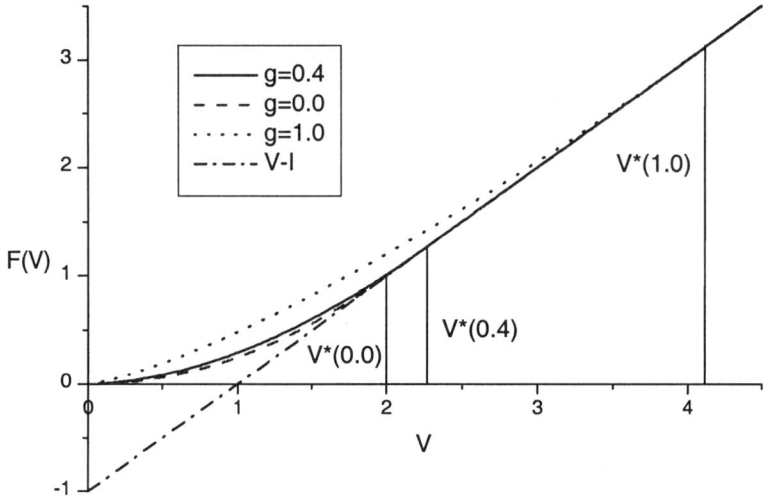

Figure 5.3: Value of the shared option to invest when competitive risk is market priced (model 1) with base case parameters $r = 0.04$, $\delta = 0.04$, $\sigma = 0.2$, $\lambda = 0.1$ and $I = 1$ for $g = 0$ (no competition), $g = 0.4$ and $g = 1.0$ (complete competitive erosion) with trigger values $V_1^*(g = 0.0) = 2.0$, $V_1^*(0.4) = 2.2649$ and $V_1^*(1.0) = 4.1583$

Since the spanning asset does not take care of the risk of competitive entry, the Poisson component is now missing. The overall uncertainty of the market development is captured by the Brownian motion. It is now easy to verify that the price dynamics of the spanning asset in the risk neutral world follows

$$dM_t = r\, M_t\, dt + \sigma\, M_t\, d\tilde{B}_t. \tag{5.10}$$

The problem is that markets do not assign a risk premium to the Poisson component of V. Therefore markets are incomplete and the equivalent martingale measure is not unique. Which measure is chosen for valuation is now a question of each

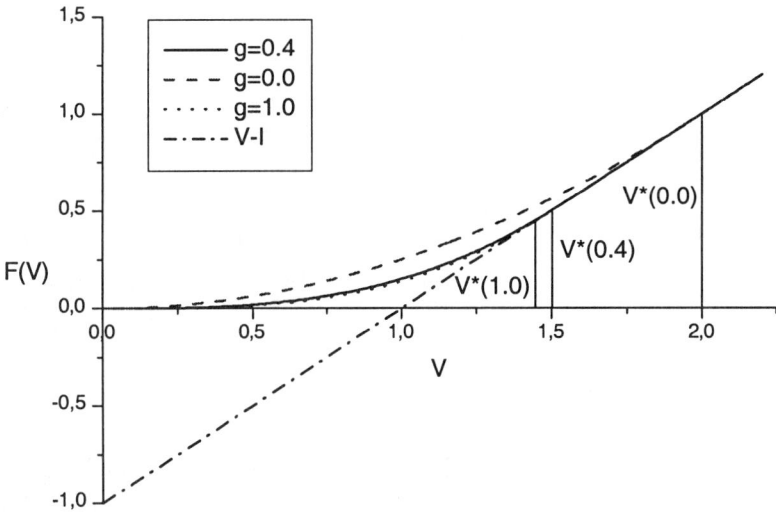

Figure 5.4: Value of the shared option to invest when competitive risk is private (model 2) with base case parameters $r = 0.04, \delta = 0.04, \sigma = 0.2, \lambda = 0.1$ and $I = 1$ for $g = 0$ (no competition), $g = 0.4$ and $g = 1.0$ (complete competitive erosion) with trigger values $V_2^*(g = 0.0) = 2.0$, $V_2^*(0.4) = 1.5046$ and $V_2^*(1.0) = 1.4561$

individual investor's risk preferences towards the Poisson risk. Without knowing the utility function of each individual investor it is not possible to select one of infinitely many measures. For simplicity it is often assumed that investors are risk neutral towards this additional source of risk.[7] Under this assumption the project value V available to the firm follows the stochastic differential equation

$$dV_t = (r - \delta)V_t dt + \sigma V_t d\tilde{B}_t - g dq_t \qquad (5.11)$$

under the selected equivalent martingale measure $\tilde{\mathbb{P}}^2$.[8]

For both models ($i = 1, 2$) the solution to the generalized timing option has to be found. Let $F_i(V)$ be the value of the inactive firm. Then the optimization

---

[7] See e.g., Amram/Kulatilaka [7].

[8] The question of how to treat private risk in real option valuation will be discussed in more detail below. The main problem to deal with is which risk is market priced and which is private to the company. The answer depends, for example, on whether spanning assets are available that contain similar selections of private and market risks as the firm or project to be valued.

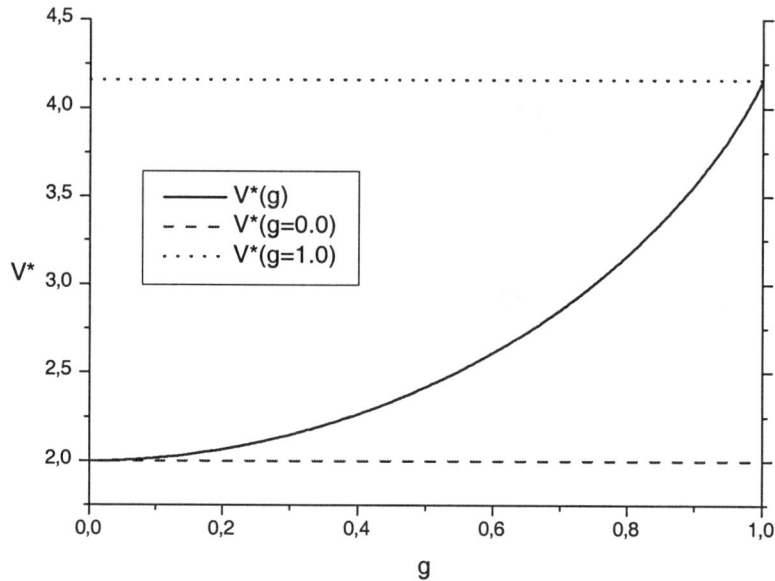

Figure 5.5: Impact of changing value erosion ratio g on V_1^* (model 1)

problem of the perpetual shared option to invest is simply given by [9]

$$F_i(V_0) = \sup_{0 \le \tau < \infty} \tilde{\mathbb{E}}^i \left[e^{-rt} (V_\tau - I) \right].$$ (5.12)

Employing the same arguments as for the timing and intensity option and using the generalized Itô formula of Section 3.2.2 leads in the case of model 1 to the ODE

$$\frac{1}{2}\sigma^2 V^2 \frac{d^2 F_1(V)}{dV^2} + (r + \lambda g - \delta)V \frac{dF_1(V)}{dV}$$
$$- (r + \lambda)F_1(V) + \lambda F_1((1-g)V) = 0$$ (5.13)

whose solution is given by

$$F_1(V) = A_1 V^{\beta_1}.$$ (5.14)

β_1 is the positive root of the characteristic equation

$$\frac{1}{2}\sigma^2 \beta_1(\beta_1 - 1) + (r + \lambda g - \delta)\beta_1 - (r + \lambda) + \lambda(1-g)^{\beta_1} = 0$$ (5.15)

---

[9]See Mordecki [173] for optimal stopping of a diffusion with jumps including the necessary proofs; see also Zein [234] for pricing of American options in a jump-diffusion model.

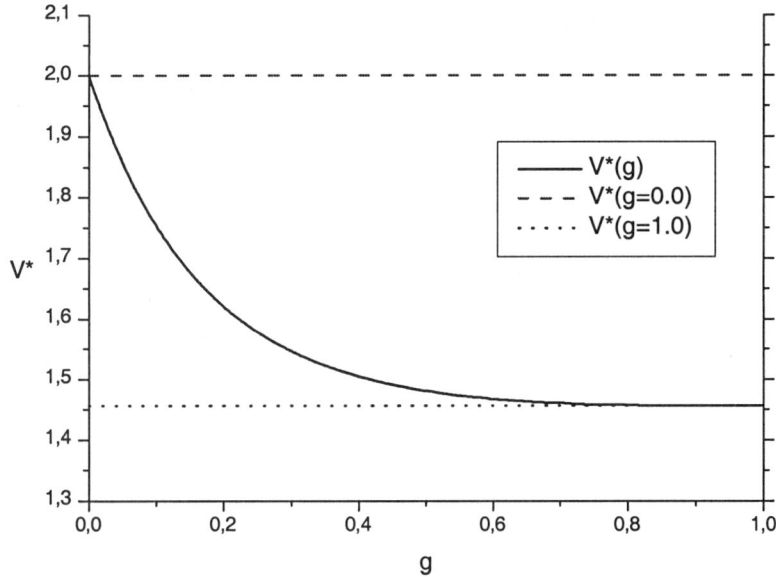

Figure 5.6: Impact of changing value erosion ratio g on V_2^* (model 2)

which can be verified by substituting (5.14) into (5.13). Unfortunately, closed form solutions for β_1 are only available for the extreme cases of no competition ($g = 0$) and complete market erosion through competitive entry ($g = 1$). For all intermediate values of g, β_1 has to be solved for numerically. The yet to determined constant A_1 and the trigger value V_1^* can be derived by the value matching and smooth pasting condition which correspond to a stochastically C^2 value function $F_1(V)$ of the generalized timing option

$$F_1(V_1^*) \;=\; V_1^* - I, \tag{5.16}$$
$$F_1'(V_1^*) \;=\; 1. \tag{5.17}$$

Thus, we find the solution to model 1 with

$$V_1^* \;=\; \frac{\beta_1}{\beta_1 - 1}\, I, \tag{5.18}$$

$$F_1(V) \;=\; \frac{I}{\beta_1 - 1}\left(\frac{V}{V_1^*}\right)^{\beta_1}. \tag{5.19}$$

The same procedure can be used to determine the ODE for model 2:

$$\frac{1}{2}\sigma^2 V^2 \frac{d^2 F_2(V)}{dV^2} + (r - \delta)V \frac{dF_2(V)}{dV}$$
$$-(r + \lambda)F_2(V) + \lambda F_2((1 - g)V) \;=\; 0 \tag{5.20}$$

with solution

$$F_2(V) = A_2 \, V^{\beta_2} \tag{5.21}$$

and β_2 the positive root of the characteristic equation

$$\frac{1}{2}\sigma^2 \beta_2(\beta_2 - 1) + (r - \delta)\beta_2 - (r + \lambda) + \lambda(1 - g)^{\beta_2} = 0. \tag{5.22}$$

The solutions for V_2^* and $F_2(V)$ of model 2 are then given by equations (5.18) and (5.19) with the subscripts replaced by 2.

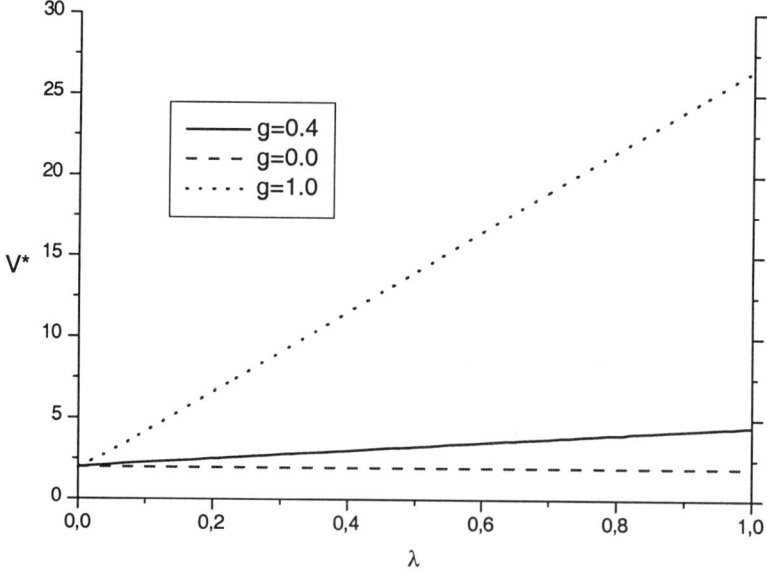

Figure 5.7: Impact of changing arrival rate λ on $V_1^*(g)$ (model 1)

It is easy to see that the only difference between both models is in the positive roots of the characteristic equations β_1 and β_2, respectively. For the base case

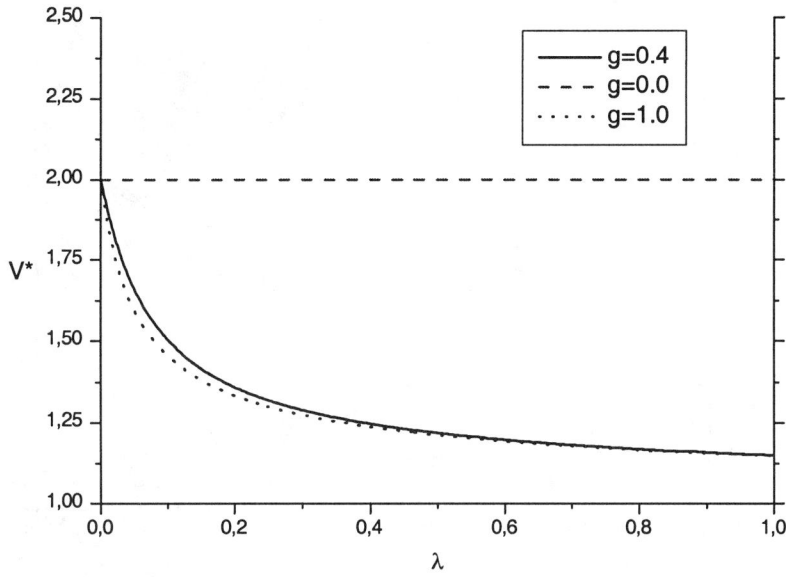

Figure 5.8: Impact of changing arrival rate λ on $V_2^*(g)$ (model 2)

parameters, $r = 0.04$, $\delta = 0.04$, $\sigma = 0.2$, $\lambda = 0.1$, $g = 0.4$ and $I = 1$, the value functions and trigger values of both models are displayed in Figures 5.3 and 5.4 along with the extreme cases of no competition ($g = 0$) and complete competitive erosion ($g = 1.0$). The results are amazing. Model 1 produces the counterintuitive result that more competition (higher g) leads to higher option values and higher trigger values $V_1^*(g)$. Therefore if the market is willing to pay a risk premium for the competitive uncertainty the firm faces, the inactive firm is worth more and has an incentive to wait even longer than in the no competition case. Consequently, there is no need to preempt other competitors. From that point of view the shared option to invest exhibits the same properties as the exclusive option to invest but with a higher volatility σ. The classical investment uncertainty relationship, that higher uncertainty raises option values but discourages investment, holds. In addition to the incentive to wait as the overall market develops, there is an incentive to learn about competitors´ actions while being protected against losses. A quite unrealistic scenario.

However, the assumption that investors assign a risk premium to competitive risk is highly questionable. On the contrary, it is more likely that competitive threat is private to the firm and therefore gives rise to model 2. The role of private risk

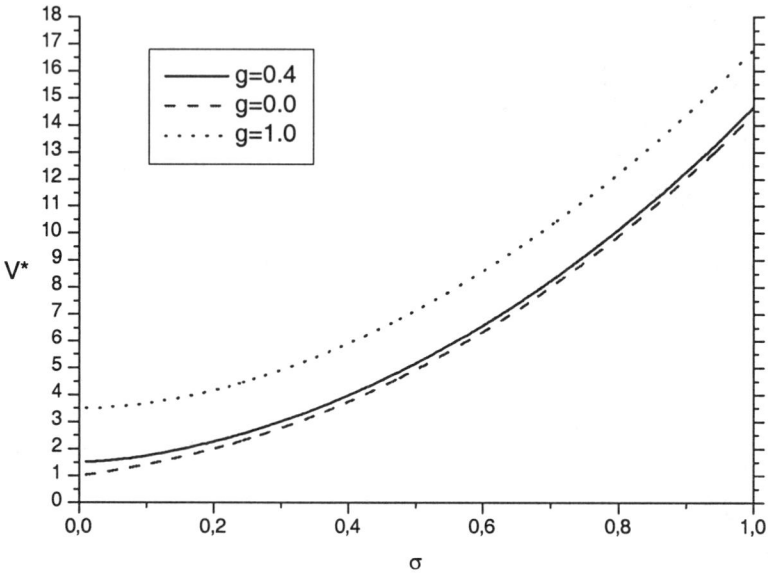

Figure 5.9: Impact of changing volatility σ on $V_1^*(g)$ (model 1)

in option values and optimal decision strategies is completely different. As can be seen in Figure 5.4, if competitive actions are interpreted as private risk, option values as well as trigger values decrease with higher possible loss by competition g. This can be explained as follows. The value of waiting is partially offset by the threat of a loss due to competitive entry and the opportunity cost of waiting therefore rises. As a result the market value V_2^* for which investment is triggered is the lower the more intense competition is. Therefore preempting other competitors is a vital strategy for the firm. By early exercise of its investment option it may have early mover advantages and earn monopoly profits until other competitors enter.

The reverse relationship between model 1 and model 2 also emerges in the sensitivity analysis of the models with respect to its input parameters. For example, the impact of changing competitive erosion ratios g is displayed in Figures 5.5 and 5.6. As expected, the entry market value V_i^* increases in g for model 1 and decreases for model 2, underlining the above described effects.

A similar effect can be observed for the mean arrival rate of competitors λ. For model 1 the trigger market values decrease linearly with λ. This is intuitive since higher λ means higher market priced risk which delays investment in model 1. The opposite result can be found for model 2 because an increase in private

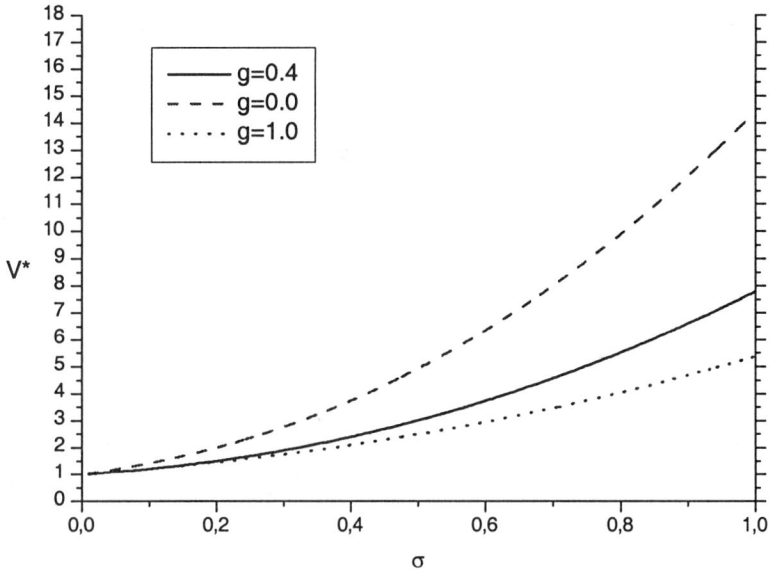

Figure 5.10: Impact of changing volatility σ on $V_2^*(g)$ (model 2)

risk erodes the option value to invest, leading to earlier investment than with no competition. Yet, there is an additional interesting result contained in model 2. Although the mean arrival rate $\lambda = 0.1$ is quite low in the base case, meaning that on average a competitor only arrives each ten years, the difference in trigger market values between complete erosion and only 40% erosion are very small. As the mean arrival rate λ increases to perhaps more realistic levels of about 1.0, the competitive erosion level seems to have almost no influence on the trigger market value. This is interesting because one would assume that the extent of competitive erosion is equally important as the arrival of competition. However, we found that the optimal exercise strategy is quite robust with respect to changes of the expected impact of competition when it occurs. On the other hand correct information about the intensity of investment λ seems to be crucial in order to be able to choose the optimal exercise strategy.

The impact of changes in volatility σ are as expected. For both models an increase in σ causes higher trigger market values because the risk represented by the Wiener process is assumed to be market priced. An increase in market priced risk always increases option values and delays investment. Model 2 shows the same pattern but with the reversed order of trigger market values again showing the optimality of a preemption strategy for a firm facing competition as a private risk. The comparative statics with respect to r and δ are similar.

These findings can be generalized to the analysis of private versus market risk. More market priced risk does encourage learning due to the possibility of higher outcomes, while the option to wait limits the downside risk. Therefore, option values as well as trigger values rise. The situation is completely different with private risk. Since private risk is not taken account of in capital markets, it represents risk that has to be borne by the firm alone. It reduces option values and consequently the value of the whole firm. In the special case considered here it reduces the investment trigger and leads to earlier investment.

The two models that have been discussed as examples for exogenous competition have demonstrated several things. First, it has to be clarified whether the risk associated with competition is private or market priced. In most cases it will be private because investors can diversify this source of risk by purchasing the stocks of direct potential competitors. However, this point is not obvious in any situation. There is for example good reason to believe that Poisson jumps in stock prices are not diversifiable completely.[10] Depending on the nature of competitive uncertainty, different strategies are optimal for the firm. In case of private risk the firm is best off by applying a preemption strategy as in the deterministic model by Trigeorgis.[11] Second, the application of Poisson processes to model competition has been demonstrated. Interestingly, the accurate determination of the intensity of competition λ seemed to be more important than the identification of the specific impact of competitive entry g. However, the question whether this effect also holds for other real options, e.g., the timing and intensity option, is yet to be answered.

5.1.2   Endogenous Competition

So far we have concentrated on real options that are either exclusive or shared with exogenous competition. The first case deals with a monopolistic market structure while the second corresponds to perfectly competitive markets. Certainly in many situations the market structure a decision maker faces will be rather oligopolistic than monopolistic or perfectly competitive, respectively.[12] The distinctive feature of oligopolistic markets is that investment decisions are contingent upon and sensitive to competitors' actions. The nature of competition is no longer exogenous but endogenous and the strategic interaction between competing firms has to analyzed. In this case a more complex combination of real option analysis and a game theoretic treatment of the situation becomes necessary. As for exogenous competition we will discuss the subject by means of an example proceeding in three main steps. First, the ideas of competitive interaction when pricing real options will be addressed. The economic interpretation will be explained as well as the

---

[10]See e.g., Kim/Oh/Brooks [113].

[11]See Trigeorgis [221, 225].

[12]See e.g., Smit/Ankum [208].

integration into the stochastic control framework and the graphical decomposition method. Lastly, the integration of game theory into the real option approach will be discussed along with some extensions for future research.[13]

Next we start with an introduction of the stylized example. Consider a company A that has developed a new product and needs to decide about its future strategy. Firm A faces uncertainty about the future development of the overall market size. It has to decide if and when it enters the market and produces the new product. In addition, once having entered firm A may also abandon its investment and exit the market if overall demand turns unfavorable. Besides the firm there is a competitor B that has developed a similar product that could be introduced in the same market as a close substitute. There are no other firms offering similar products — at least not in the near future. Therefore the competitive situation is a duopoly. The difference between the two firms involved is in their production capabilities. Firm A is assumed to have a competitive advantage due to a superior production technology that allows for producing at lower variable cost than firm B. Because the market price of the new product is sensitive with respect to the supply offered by the two firms, the problem for both firms is when to enter and exit the market and which quantities to offer. For simplicity it is assumed that once a firm enters and chooses its quantity (capacity) Q_i, it remains fixed as long as the firm is operating. Furthermore we assume that capital markets and both competitors have perfect information about each others' cost structure.

The output price of the product is determined by the stochastic shifts in the demand curve θ and the total quantity Q supplied by both firms. The inverse demand curve fulfills the linear relationship

$$P_t = \Theta_t - \lambda Q_t \tag{5.23}$$

where the stochastic demand shift parameter can be interpreted as the maximal demand when supply is zero. The slope λ is positive and constant. Changes in the demand parameter Θ can be due to changes in consumer tastes, natural forces, political developments, or the business cycle. For tractability the demand parameter is assumed to follow a geometric Brownian motion under the equivalent martingale measure[14] $\tilde{\mathbb{P}}$ according to

$$d\Theta_t = \mu \, \Theta_t \, dt + \sigma \, \Theta_t \, d\tilde{B}_t, \qquad \Theta_0 \text{ given.} \tag{5.24}$$

The parameters μ and σ are constant and positive. By specifying the inverse demand function in the given way, price changes are endogenized and change due to quantity

---

[13]An in-depth treatment of game theory would be beyond the subject of this book and will therefore be kept on an expository level. Moreover, we are more concerned with the economic intuition behind an integration of game theory into real option analysis.

[14]It might be cumbersome if not even impossible to derive the equivalent martingale measure for an artificial variable like maximal market demand. To our knowledge there has been no empirical study on that subject so far. However, with a day by day increasing number of traded contracts there is a good chance to find a proxy for the market demand parameter Θ at capital markets.

adjustments by both firms. Intuitively, as the demand parameter Θ increases, the market demand for the product increases. For a given quantity supplied, the product price will increase by the same amount as the demand parameter. At the entry and exit points, the product price will have a discrete downward or upward jump due to the change of total quantity supplied. The magnitude of the jump is determined by λ. Notice that when λ is equal to zero, the market price is exogenous and follows the same stochastic process as Θ.

As for all above discussed models there is no building time (or time lags) considered.[15] Hence, the firm can get the production capacity on stream immediately after having decided to enter the market. Since in that instance the demand parameter does not change, the price is adjusted immediately by a downward jump.

Each firm selects the demand parameter levels at which it enters and leaves the market and the production quantity to maximize the value of the firm, assuming that the other firm will do the same. This is the standard treatment of oligopolistic competition and is equivalent to the Nash equilibrium concept in dynamic game theory. By assuming that both firms know which firm has a cost advantage, the two players are prevented from making simultaneous moves, which simplifies the game theoretic analysis. In case of no information about the other firm's cost structure, neither firm can condition its strategy on the other firm's decision and game theory would not be applicable. In addition, if both firms have equal costs and perfect information, we would still have simultaneous moves. The assumption of cost differences and perfect information might be a first reasonable approximation of reality.[16]

The firm will enter the market only when the output price is well above the unit cost of production and exit the market only when the price drops below the unit cost of production adjusted for the opportunity cost of entry and exit. These hysteresis effects are caused by the option values generated by irreversible entry and exit costs under demand uncertainty. Consequently, both firms will have an entry trigger demand level and a lower exit trigger demand level. This strategy occurs as well in the monopolistic setting as in the duopoly. The question to be answered is how competition alters the optimal investment strategies of the two firms regarding its entry, exit and capacity decisions. Firms usually have a different set of capabilities and competitive advantages which makes them adopt different strategies. The model takes this into account by specifying different variable costs of production, all other things being equal. Other specifications of competitive advantages are, of course, possible. In order to determine the influence of competition on the decision strategies we show how to model the monopolistic case and the duopolistic case in the stochastic control framework. The duopolistic case will be further distinguished by the kind of equilibrium the firms may attain. Namely, the option game results in a Cournot equilibrium when both firms maximize their profits simultaneously

---

[15]This assumption will be relaxed in the next section.

[16]Asymmetric (or incomplete) information option games have been treated by Lambrecht/Perraudin [135].

while treating the decision variables of the competitor as fixed parameters. If on the other hand the low cost firm that has the competitive advantage acts as a leader, it will anticipate the optimal strategy of the high cost firm which acts as a follower and will optimize its value over all possible optimal strategies of the follower. This case results in a Stackelberg leader-follower equilibrium.

Similar to the entry and exit model of Section 4.2 the monopolistic case can be represented by an impulse control problem (or generalized switching option) with two states — inactive and operating with capacity Q_A — between which reversible switches may take place. The time horizon for the problem considered here is infinity and the underlying source of uncertainty is given by the stochastic demand parameter Θ. The action space $\mathcal{Z}$ consists of the states inactive (0) and operating with capacity Q_A. Notice that by choosing the capacity Q_A, the model is the reversible switching analogy of the irreversible timing and intensity option of Section 4.3. The switching cost function is given by the entry cost

$$H(t, \Theta_t, 0, Q_a) = m \, Q_A^a, \qquad m > 0, a \geq 1 \qquad (5.25)$$

and the exit cost

$$H(t, \Theta_t, Q_a, 0) = E. \qquad (5.26)$$

The parameter a again represents the assumption that additional capacity becomes more and more expensive. The exit cost E may be positive, negative or zero depending on the specific situation. While operating, the monopolistic firm earns an instantaneous cash flow of revenues generated by supply Q_A minus production cost. This can be expressed as

$$f(t, \Theta_t, Z_t) \;=\; f(Z_t) \qquad (5.27)$$
$$=\; \begin{cases} \Theta_t Q_A - \lambda Q_A^2 - c_A Q_A & \text{for } Z_t = Q_A, \\ 0 & \text{for } Z_t = 0, \end{cases}$$

where the running production costs c_A of Firm A are assumed to be constant.

The graphical representation of this generalized switching option is displayed in Figure 5.11 where it was taken account for the intensity option to choose capacity Q_A by writing Q in brackets.

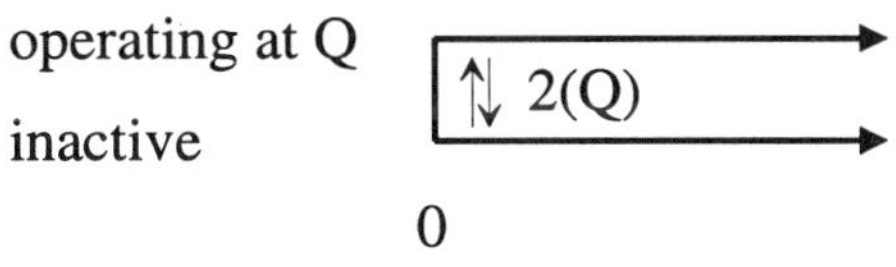

Figure 5.11: Entry and exit decision in monopoly with variable capacity Q_A

In order to value this generalized switching option we have to formulate the corresponding impulse control problem. A feasible impulse control strategy in the monopolistic case consists of a sequence of stopping times θ_i and corresponding impulse control interventions $\zeta_i = 0$ or $\zeta_{i+1} = Q_A$. Let $w \in \mathcal{W}$ denote an impulse control (entry and exit times and actions of the monopolistic firm). Then the impulse control optimization problem to solve is (assuming that we start in state inactive)

$$V^A(0, \Theta_0, 0) = \sup_{w \in \mathcal{W}} \tilde{\mathbb{E}} \left[\int_0^\infty e^{-rt} f(t, \Theta_t, Z_t) \, dt \right.$$

$$\left. - \sum_{i=1}^\infty e^{-r\theta_i} H(\theta_i, \Theta_{\theta_i}, \zeta_{i-1}, \zeta_i) \right]. \qquad (5.28)$$

While a solution to this optimization problem is hard to derive, the value function can equivalently be obtained by solving the corresponding quasi-variational inequalities. For the above problem we formulate the maximum operator

$$\mathcal{M}V^A(t, \Theta_t, z) = \max_{\zeta \in \mathcal{Z}\backslash\{z\}} \left\{ V^A(t, \Theta_t, \zeta) - H(t, \Theta_t, z, \zeta) \right\}. \qquad (5.29)$$

The maximum operator distinguishes the two cases where the firm is inactive or operating. The inactive firm ($z = 0$) maximizes at entry over all possible capacity levels ($\zeta = Q_A$). The operating firm ($z = Q_A$) can only switch to state inactive ($\zeta = 0$) so that there is strictly speaking no optimization.

Employing Itô's lemma as in Section 3.2.2 yields the quasi-variational inequalities for the value of the monopolistic firm:

$$\frac{1}{2}\sigma^2\Theta^2\frac{d^2V^A}{d\Theta^2} + \mu\Theta\frac{dV^A}{d\Theta} - rV^A + f(Z_t) \leq 0$$

$$V^A \geq \mathcal{M}V^A. \qquad (5.30)$$

One of the inequalities is an equality.

In order to solve this system of quasi-variational inequalities we start by considering state inactive first. If the firm is inactive it may stay inactive or switch to the operating state and choose the optimal capacity level. It is easy to see that this is exactly the same situation as for the timing and intensity option except that the value of the operating firm now contains an additional option value for later exit of the market. Let V_0^A and V_1^A be the value of firm A being inactive or operating with capacity Q_A, respectively. The solution V_0^A can be found by solving the ODE

$$\frac{1}{2}\sigma^2\Theta^2\frac{d^2V^A}{d\Theta^2} + \mu\Theta\frac{dV^A}{d\Theta} - rV^A = 0. \qquad (5.31)$$

Then the value of the inactive firm is given as

$$V_0^A(\Theta) = A\,\Theta^{\beta_1} \qquad (5.32)$$

which consists solely of the option value to enter the market. β_1 is the positive root of the characteristic equation

$$\frac{1}{2}\sigma^2\beta(\beta - 1) + \mu\beta - r = 0. \qquad (5.33)$$

Consequently, the roots are

$$\beta_{1/2} = \frac{1}{2} - \frac{\mu}{\sigma^2} \pm \sqrt{\left(\frac{\mu}{\sigma^2} - \frac{1}{2}\right)^2 + \frac{2r}{\sigma^2}}. \tag{5.34}$$

Similarly, the value of the operating firm V_1^A satisfies the ODE

$$\frac{1}{2}\sigma^2\Theta^2\frac{d^2V^A}{d\Theta^2} + \mu\Theta\frac{dV^A}{d\Theta} - rV^A + \Theta_t Q_A - \lambda Q_A^2 - c_A Q_A = 0. \tag{5.35}$$

Its solution is given by

$$V_1^A(\Theta) = Q_A\left(\frac{\Theta}{r-\mu} - \frac{\lambda Q_A + c_A}{r}\right) + B\,\Theta^{\beta_2} \tag{5.36}$$

where β_2 is the negative root given by (5.34). The first term on the right side of equation (5.36) reflects the value generated by the cash flows of the operating firm while the second term describes the option value of exiting the market.

So far, we have solved for the values of the inactive and operating firm having used the information of the quasi-variational inequalities, so that in the continuation region the first inequality for both the inactive and operating firm has to be an equality. Next the maximum operator is used to determine the trigger values of Θ for which switching between the two states takes place. Let us start with exiting the market because there is only one possible state to consider. At the time instant when it is optimal to exit the market, we must have that the value of the inactive firm minus exit cost E is at least as valuable as keeping the firm operating. Therefore at the point of indifference Θ_l we must have

$$V_1^A(\Theta_l) = V_0^A(\Theta_l) - E. \tag{5.37}$$

Moreover, we need V^A to be stochastically C^2 which means that V^A is at least once continuously differentiable at Θ_l,

$$V_1^{A'}(\Theta_l) = V_0^{A'}(\Theta_l). \tag{5.38}$$

The value of the market demand parameter for which the firm is indifferent between staying inactive and starting operation is denoted by Θ_h. The same arguments apply as above with the extension that we have to take care of the optimally chosen capacity level Q_A. As for the timing and intensity option, we solve the problem by separating the timing and intensity effect and assuming first that Q_A is fixed. This gives the two equations

$$V_0^A(\Theta_h) = V_1^A(\Theta_h) - mQ_A^a, \tag{5.39}$$
$$V_0^{A'}(\Theta_h) = V_1^{A'}(\Theta_h). \tag{5.40}$$

Equations (5.37) to (5.40) are the familiar value matching and smooth pasting conditions for optimal exercise of the entry and exit options. Plugging the general solutions of V_0^A and V_1^A of equations (5.32) and (5.36) into (5.37) to (5.40) yields a system of four equations for the four unknown variables A, B, Θ_l and Θ_h which can be solved numerically:[17]

$$A\Theta_l^{\beta_1} - E = Q_A \left(\frac{\Theta_l}{r-\mu} - \frac{\lambda Q_A + c_A}{r} \right) + B\Theta_l^{\beta_2}, \qquad (5.41)$$

$$A\beta_1\Theta_l^{\beta_1-1} = Q_A \frac{1}{r-\mu} + B\beta_2\Theta_l^{\beta_2-1}, \qquad (5.42)$$

$$A\Theta_h^{\beta_1} = Q_A \left(\frac{\Theta_h}{r-\mu} - \frac{\lambda Q_A + c_A}{r} \right) + B\Theta_h^{\beta_2} - mQ_A^a, \quad (5.43)$$

$$A\beta_1\Theta_h^{\beta_1-1} = Q_A \frac{1}{r-\mu} + B\beta_2\Theta_h^{\beta_2-1}. \qquad (5.44)$$

Besides this, the firm must also select the optimal production quantity Q_A according to the maximum operator of (5.31). It is assumed that as for the timing and intensity option above the firm chooses Q_A to maximize V_0^A. Therefore at the entry point the firm either maximizes $V_0^A(\Theta_h)$ or equivalently $V_1^A(\Theta_h) - mQ_A^a$. From equation (5.32) it is clear that maximizing V_0^A with respect to Q_A is equivalent to maximizing A with respect to Q_A. Formally, the capacity choice problem is therefore to maximize A with respect to Q_A subject to the equations (5.41) to (5.44). Different methods can be used to solve this maximization problem. For example, when using a Lagrange multiplier method the Lagrangian L is defined as

$$L = A + \eta_1 EQ5.41 + \eta_2 EQ5.42 + \eta_3 EQ5.43 + \eta_4 EQ5.44 \qquad (5.45)$$

where EQ denotes equation. Setting the derivatives of L with respect to the nine unknowns A, B, Θ_l, Θ_h, Q_A and η_1 to η_4 equal to zero and solving the resulting nine nonlinear equations numerically yields the desired result for the monopolistic firm.

We will not discuss numerical examples of this model in detail but will rather concentrate on the main results of the monopolistic case. First, it is evident that for all reasonable parameter values uncertainty drives a wedge between the entry and exit decision. As a consequence the entry trigger when the firm is idle, Θ_h, is well above the average production cost which consists of the variable cost plus the cost of installed capital. The average production cost forms the so-called *Marshallian trigger*[18] which is the trigger under certainty. Vice versa, uncertainty lowers the exit trigger Θ_l so that a higher chance of imminent recovery is taken into account. Θ_l in contrast is well below the Marshallian trigger Θ_M. Thus, we have the general relationship between trigger levels

$$\Theta_l < \Theta_M < \Theta_h \qquad (5.46)$$

---

[17] A numerical analysis of a similar model for fixed capacity Q_A is carried out in Dixit/Pindyck [72], p. 223, along with an in-depth discussion of its microeconomic consequences.

[18] See e.g., Abel [1].

for all values $\sigma > 0$. This establishes the hysteresis effect.

Notice that the separation of the timing and intensity decision allowed us to specify conditions for reducing a time homogeneous impulse control problem to the numerical solution of several nonlinear equations. However, even for the time homogeneous problem this is a rare exception. None the less the general procedure, first to find the solution in the continuation region and to determine its boundaries afterwards using the maximum operator, remains valid for numerical solution techniques as well.[19]

The entry and exit decision of a monopolistic firm was the first example of how to find solutions of generalized switching options by solving impulse control problems. As we will see for the duopoly case the analysis can be easily extended to a dynamic game of impulse control problems.

Next consider the case of a two firm industry. In case of this competitive situation each of the two firms holds a generalized switching option with exactly the same states, action spaces ($\mathcal{Z}^A = \mathcal{Z}^B$) and switching cost functions ($H^A = H^B$). Thus, these generalized switching options can be described using impulse control methods. However, competition causes the two models to be interconnected. In the duopoly considered here both competitors influence each others' instantaneous cash flows by their entry/exit and capacity choice decisions. For that reason the two impulse control problems have to be solved for simultaneously. Since each firm can be in one of two states, the whole market has to be in one of four different states. As it will turn out the optimally chosen strategies of both competitors may rule out some of these states as not viable. To start with we work out the cross dependence of the models by specifying the cash flows of firms A and B in all four possible states of the system. This will give us a means to find a graphical representation of the described strategic game. One could proceed in the same way as for the monopolistic case. Both firms are in one of two states, either inactive $Z_t^i = 0$ ($i = A, B$) or operating with capacity $Z_t^A = Q_A$ or $Z_t^B = Q_B$, respectively. Since each firm wants to maximize its value we have to take account of the interaction of the entry and exit strategies of both firms simultaneously. The interaction between the two firms is displayed by the influence their strategies have on the competitor's operating cash flow. All other parameters of the impulse control models, like the switching cost functions are independent of competition and are given as in the monopolistic case. Considering the cash flows of firm A if it is operating, we have two cases to distinguish. The first is when firm B has not yet entered the market and the overall supply is Q_A. Second, after firm B has entered the market the overall supply is $Q = Q_A + Q_B$ leading to a reduction in price and consequently to a reduction of the instantaneous cash flow of firm A. Similarly, if firm B is first in the market it will earn monopoly profits with total supply Q_B until firm A enters and the overall supply jumps to $Q = Q_A + Q_B$. Lastly, if none of the firms is operating ($Z_t^A = Z_t^B = 0$) the cash flows of both firms are zero. The cash flows $f_i(Z_t^A, Z_t^B)$ earned by each firm in the four cases are displayed in Table 5.1.

---

[19]See the case study presented in Chapter 6.

No.	States		Cash Flows	
	Z_t^A	Z_t^B	$f_A(Z_t^A, Z_t^B)$	$f_B(Z_t^A, Z_t^B)$
1	0	0	0	0
2	Q_A	0	$\Theta Q_A - \lambda Q_A^2 - c_A Q_A$	0
3	0	Q_B	0	$\Theta Q_B - \lambda Q_B^2 - c_B Q_B$
4	Q_A	Q_B	$\Theta Q_A - c_A Q_A$ $-\lambda Q_A(Q_A + Q_B)$	$\Theta Q_B - c_B Q_B$ $-\lambda Q_B(Q_A + Q_B)$

Table 5.1: Cash flow earned by firm i in each of the four possible states.

Now that both models are sufficiently specified there are several ways conceivable to represent the strategic game in graphical form. To keep the graphs as simple as possible for means of communication, we have chosen a similar symbol as in the monopolistic case. Since we take an individual firm's view, the modelling element displayed in Figure 5.12 only represents the generalized switching option of firm A. It has the opportunity to switch between two states with an additional intensity option to choose Q_A which is displayed by $2(Q_A)$. Similarly, its competitor has the same means of flexibility denoted by $2(Q_B)$. That the overall market situation is a duopoly is indicated by the word 'game'.

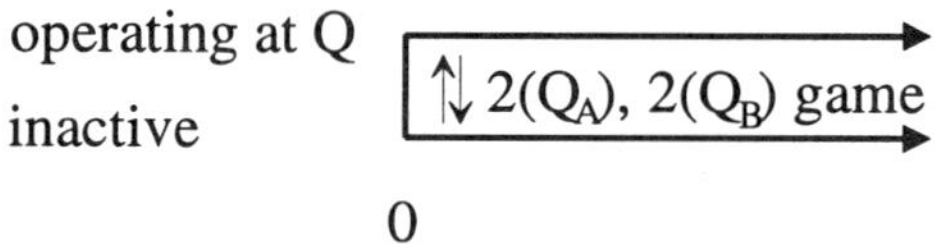

Figure 5.12: Entry and exit decision in duopoly with variable capacities Q_A and Q_B

The next step is to solve the duopoly impulse control game. Similar to the monopolistic case the solution boils down to finding the entry and exit triggers for both firms as well as the corresponding optimal capacity levels. However, the situation gets more complicated because it is not a priori clear which firm will enter first and which last. For that reason it is assumed that because firm A has a lower variable cost of production ($c_A < c_B$) firm A will enter earlier and exit later than firm B.[20] Furthermore, we rule out simultaneous entry or mixed strategies. Thus, firm A will act as the leader and firm B as the follower. Since firm A enters earlier and exits later than firm B, we conjecture that the trigger levels fulfill the

---

[20]Actually, it could be proved using impulse control techniques that it is never an optimal strategy for firm B to enter earlier or exit later than firm A for reasonable parameter values as long as the only difference between the two firms is in their variable production cost.

relationship

$$\Theta_l^A < \Theta_l^B \quad \text{and} \quad \Theta_h^A < \Theta_h^B \tag{5.47}$$

where Θ_h^i and Θ_l^i are the entry and exit demand parameters for the firms $i = A$ and $i = B$, respectively. Given (5.47), firm A can be either operating or inactive when firm B is inactive. Similarly, firm B can be either operating or inactive when firm A is operating. However, firm B must be inactive when firm A is inactive, therefore making case three above not applicable. This reduces the number of states to consider to three.

We introduce the following nomenclature for firm values. The first subscript following V refers to the state of firm A. It is 1 when firm A is operating with capacity Q_A and 0 when A is inactive. The last subscripted number refers to firm B in the same way.

No.	State		Value	
	Firm A	Firm B	Firm A	Firm B
1	0	0	V_{00}^A	V_{00}^B
2	Q_A	0	V_{10}^A	V_{10}^B
4	Q_A	Q_B	V_{11}^A	V_{11}^B

Table 5.2: Firm values for each possible state of the option game

The impulse control problem of each firm is then given by the sets of quasi-variational inequalities

$$\frac{1}{2}\sigma^2\Theta^2\frac{d^2V^A}{d\Theta^2} + \mu\Theta\frac{dV^A}{d\Theta} - rV^A + f_A(Z_t^A, Z_t^B) \leq 0$$

$$V^A \geq \mathcal{M}V^A, \tag{5.48}$$

one of the inequalities is an equality,

for firm A and

$$\frac{1}{2}\sigma^2\Theta^2\frac{d^2V^B}{d\Theta^2} + \mu\Theta\frac{dV^B}{d\Theta} - rV^B + f_B(Z_t^A, Z_t^B) \leq 0,$$

$$V^B \geq \mathcal{M}V^B, \tag{5.49}$$

one of the inequalities is an equality,

for firm B, respectively. We will shortly explain how to solve it. First consider firm A. If firm A is inactive — and so is firm B — its value is the solution to the ODE

$$\frac{1}{2}\sigma^2\Theta^2\frac{d^2V_{00}^A}{d\Theta^2} + \mu\Theta\frac{dV_{00}^A}{d\Theta} - rV_{00}^A = 0 \tag{5.50}$$

given as in the monopolistic case by[21]

$$V_{00}^A(\Theta) = A_1 \,\Theta^{\beta_1}. \tag{5.51}$$

If firm A has already invested but firm B is still inactive, the value function satisfies the ODE

$$\frac{1}{2}\sigma^2\Theta^2\frac{d^2V_{10}^A}{d\Theta^2} + \mu\Theta\frac{dV_{10}^A}{d\Theta} - rV_{10}^A + \Theta Q_A - \lambda Q_A^2 - c_A Q_A = 0 \tag{5.52}$$

with solution

$$V_{10}^A(\Theta) = A_2 \,\Theta^{\beta_1} + A_3 \,\Theta^{\beta_2} + \frac{\Theta Q_A}{r-\mu} - \frac{\lambda Q_A^2 + c_A Q_A}{r}. \tag{5.53}$$

The investment trigger Θ_h^A at which firm A enters the market and applies an impulse control is subject to the familiar value matching and smooth pasting conditions

$$V_{00}^A(\Theta_h^A) = V_{10}^A(\Theta_h^A) - m Q_A^a, \tag{5.54}$$

$$V_{00}^{A'}(\Theta_h^A) = V_{10}^{A'}(\Theta_h^A). \tag{5.55}$$

If both firms are operating, the value of firm A changes due to the different instantaneous cash flow now earned by firm A. The corresponding ODE is

$$\frac{1}{2}\sigma^2\Theta^2\frac{d^2V_{11}^A}{d\Theta^2} + \mu\Theta\frac{dV_{11}^A}{d\Theta} - rV_{11}^A + \Theta Q_A \tag{5.56}$$

$$-\lambda Q_A(Q_A + Q_B) - c_A Q_A = 0$$

with solution

$$V_{11}^A(\Theta) = A_4 \,\Theta^{\beta_2} + \frac{\Theta Q_A}{r-\mu} - \frac{\lambda Q_A(Q_A + Q_B) + c_A Q_A}{r}. \tag{5.57}$$

However, it is firm B which selects the investment trigger Θ_h^B that distinguishes V_{10}^A and V_{11}^A. Firm A remains in its operating state without applying any impulse control. Because the value function of firm A's impulse control problem is continuous, as long as no impulse control is triggered for firm A the following condition must hold at the investment trigger Θ_h^B of firm B

$$V_{10}^A(\Theta_h^B) = V_{11}^A(\Theta_h^B). \tag{5.58}$$

This additional interaction boundary condition can be interpreted economically as follows. The value of the firm should be the same before and after the competitor

---

[21]The corresponding constants of the general solution of the ODE have been eliminated using straightforward arguments about the values of the entry and exit options as the demand parameter gets small or large.

takes its action, since both firms and the market have the same correct information and hence will anticipate the entry decision by firm B.

Vice versa, if both firms are operating, firm B will exit the market earlier by assumption at the demand parameter level Θ_l^B. This results in another interaction boundary condition for firm A,

$$V_{11}^A(\Theta_l^B) = V_{10}^A(\Theta_l^B). \tag{5.59}$$

Finally, if the demand parameter further decreases to the exit threshold Θ_l^A we have the value matching and smooth pasting condition for Θ_l^A:

$$V_{10}^A(\Theta_l^A) = V_{00}^A(\Theta_l^A) - E, \tag{5.60}$$

$$V_{10}^{A'}(\Theta_l^A) = V_{00}^{A'}(\Theta_l^A). \tag{5.61}$$

This completes the discussion of the conditions that the value function V^A of the impulse control problem of firm A satisfies.

For firm B we can proceed in exactly the same way. The corresponding value functions, switching conditions and interaction boundaries when firm A takes its actions are presented in the following nine equations.

$$V_{00}^B(\Theta) = B_1 \Theta^{\beta_1} \tag{5.62}$$

$$V_{10}^B(\Theta) = B_2 \Theta^{\beta_1} + B_3 \Theta^{\beta_2} \tag{5.63}$$

$$V_{11}^B(\Theta) = B_4 \Theta^{\beta_2} + \frac{\Theta Q_B}{r - \mu} - \frac{\lambda Q_B(Q_A + Q_B) + c_B Q_B}{r} \tag{5.64}$$

$$V_{10}^B(\Theta_h^B) = V_{11}^B(\Theta_h^B) - m Q_B^a \tag{5.65}$$

$$V_{10}^{B'}(\Theta_h^B) = V_{11}^{B'}(\Theta_h^B) \tag{5.66}$$

$$V_{11}^B(\Theta_l^B) = V_{10}^B(\Theta_l^B) - E \tag{5.67}$$

$$V_{11}^{B'}(\Theta_l^B) = V_{10}^{B'}(\Theta_l^B) \tag{5.68}$$

$$V_{00}^B(\Theta_h^A) = V_{10}^B(\Theta_h^A) \tag{5.69}$$

$$V_{10}^B(\Theta_l^A) = V_{00}^B(\Theta_l^A) \tag{5.70}$$

In contrast to the usual solution of impulse control problems, conditions (5.58), (5.59), (5.69) and (5.70) capture the competitive interactions between the two firms. The values of each firm are now affected by the other firm's actions with regard to entry, exit and capacity choice. When firm A decides to enter and leave the market, its actions affect the value of firm B, and hence indirectly affect the entry, exit and capacity choice decisions of firm B and vice versa.

Now consider the firm's capacity choice problems. The capacity choice along with the entry and exit demand parameters is also a strategic variable. In the following there are two ways considered for how each firm can choose its capacity. In the Cournot setting, firm A maximizes A_1 with respect to Q_A, treating the competitor's choice Q_B as parametric, while firm B maximizes B_1 with respect to Q_B, treating firm A's choice Q_A as parametric. The solution obtained by the two simultaneous optimization problems yields the optimal entry, exit and capacity choices of the two firms. The resulting equilibrium is the classical Cournot equilibrium. Mathematically, it is the same as adding two optimization objectives. As before the parameters A_1 and B_1 are to be maximized, which is equivalent to maximizing the values of the inactive firms. The resulting joint maximization problem using again the Lagrange multiplier method under the constraint that the twelve equations (5.51), (5.54), (5.55), (5.58) to (5.60) and (5.65) to (5.70) hold is

$$\max_{Q_A} L_A = \max_{Q_A} \{ A_1 + \sum_{i=1}^{12} \eta_i^A \times \text{constraint } i \}, \qquad (5.71)$$

$$\max_{Q_B} L_B = \max_{Q_B} \{ B_1 + \sum_{i=1}^{12} \eta_i^B \times \text{constraint } i \}. \qquad (5.72)$$

Setting the derivatives of L_A with respect to A_1, A_2, A_3, A_4, B_1, B_2, B_3, B_4, Θ_l^A, Θ_h^A, Θ_l^B, Θ_h^B, Q_A and η_1^A to η_{12}^A as well as the derivatives of L_B with respect to A_1, A_2, A_3, A_4, B_1, B_2, B_3, B_4, Θ_l^A, Θ_h^A, Θ_l^B, Θ_h^B, Q_B and η_1^B to η_{12}^B to zero yields 38 equations which can be solved for the 38 unknowns. Notice that L_A is optimized with respect to Q_A but not with respect to Q_B, since for firm A, Q_B is viewed as an exogenously given parameter. The same holds true for L_B with respect to Q_A.

In the Stackelberg setting, firm A is the leader since it is the low cost firm. Firm A can choose its capacity to influence firm B's capacity choice as well as firm B's entry and exit strategies. The ability to select capacity first under competition is an important strategic advantage. For each capacity level firm A chooses, firm B will pick its own level to maximize its value. Hence, the optimal capacity for firm B, Q_B^*, is a function of firm A's capacity choice Q_A. Firm A knows with certainty the value of Q_B^* once it chooses Q_A. So instead of assuming Q_B to be parametric — the Cournot case — firm A anticipating firm B's reaction $Q_B^*(Q_A)$ will pick the capacity level that will maximize its value. The corresponding maximization problem can again be solved using the Lagrange multiplier method. Firm A has to maximize

$$\max_{Q_A} L_A = \max_{Q_A} \{ A_1 + \sum_{i=1}^{12} \eta_i^A \times \text{constraint } i \} \qquad (5.73)$$

with $Q_B^*(Q_A)$ and firm B

$$\max_{Q_B} L_B = \max_{Q_B} \{ B_1 + \sum_{i=1}^{12} \eta_i^B \times \text{constraint } i \} \qquad (5.74)$$

with Q_A treated as constant.

The hierarchical nature of the two optimization problems becomes obvious. The Lagrangian approach results in the solution of 51 equations which is somewhat messy. Alternatively, one can compute the optimal $Q_B^*(Q_A)$ for different values of Q_A from the 25 equations resulting from the Lagrangian of firm B's optimization problem alone and afterwards pick Q_A such as to maximize firm A's value.

A detailed numerical study of the presented strategic game is left for future research. However, some of the main results that can be conjectured from the model are presented here. The results of the timing and intensity option discussed in Section 4.3 can be transformed to the strategic game. The basic question to answer is how the firms' entry and exit demand levels, production quantities and values behave in the case of no competition (i.e., monopoly) and in the case of competition (i.e., Cournot or Stackelberg equilibrium). For the monopoly firm one would expect that the low cost firm A enters earlier and exits later than in the case where the high cost firm B is a monopolist. Firm A can afford to do this because of its low production cost. Firm A can make money at a lower price than firm B. Consequently, firm A enters at a lower demand parameter and exits at a lower exit demand parameter. Similar to the findings in Section 4.3 for the timing and intensity option, the optimal capacity Q_A for the low cost firm A should be lower than the optimal capacity choice Q_B of the high cost firm B, admitting the same co-movement of higher entry trigger and higher capacity. Lastly, the value of firm A is obviously higher than the value of firm B.

Now consider the case where the firms are Cournot competitors. As a duopolist it is clear that firm A will enter earlier and exit later in comparison to the case where it is a monopolist. Furthermore, firm A will choose a lower capacity as in the monopolistic case to reap the benefit of being the sole player in the market. By way of contrast, the high cost firm will enter considerably later and leave earlier relative to the case where it is a monopolist. Since the low cost firm is already operating when the high cost firm enters the market, some of the demand has already been picked up by firm A. Therefore, firm B has to wait for the demand to go up before it is profitable to enter the market. Moreover, with a higher entry demand parameter, firm B should be expected to select a higher capacity Q_A as in the monopolistic setting due to the observable co-movement of timing and intensity of investment. Finally, both firm values are smaller in comparison to the cases where they are monopolists.

In the case of Stackelberg competition the low cost firm is the leader while the high cost firm is the follower. Under Cournot competition each firm chooses production quantity to maximize its value, treating the other firm's production quantity as parametric. In other words, $Q_A(Q_B)$ is a function of Q_B and $Q_B(Q_A)$ is a function of Q_A. The equilibrium is obtained by solving these two reaction functions. Thus, neither firm chooses its capacity strategically. Consequently, the low cost firm does not exploit its strategic advantage by utilizing its leader role when determining production capacity. However, under Stackelberg competition, the low cost firm uses its capacity choice to its full strategic advantage. Firm B chooses its production quantity Q_B as a function of Q_A, but firm A chooses Q_A anticipating firm B's optimal choice $Q_B^*(Q_A)$. Firm A is in control of the only variable in the system, Q_A. Once it picks Q_A the other remaining variables can

	Firm A	Firm B
Entry demand level	$\Theta_h^C \leq \Theta_h^{St} \leq \Theta_h^M$	$\Theta_h^M \leq \Theta_h^C \leq \Theta_h^{St}$
Exit demand level	$\Theta_l^C \leq \Theta_l^{St} \leq \Theta_l^M$	$\Theta_l^M \leq \Theta_l^C \leq \Theta_l^{St}$
Capacity choice	$Q_A^C \leq Q_A^{St} \leq Q_A^M$	$Q_B^M \leq Q_B^C \leq Q_B^{St}$
Firm value	$V_A^C \leq V_A^{St} \leq V_A^M$	$V_B^{St} \leq V_B^C \leq Q_B^M$

Table 5.3: Qualitative results of strategies and firm values under monopoly (M), Cournot (C) and Stackelberg (St) competition

be determined by firm B's optimization problem. Consequently, under Stackelberg competition the low cost firm A holds the *strategic option to lead*. If it finds out that it is not preferable to lead, it could just behave as Cournot duopolist, assuming the high cost firm would not react to its action. Hence, in the worst case, the value of the low cost firm as a leader will be at least as high as its value not acting strategically which is the Cournot case.

This logic directly points to the qualitative nature of the optimal entry, exit and capacity choice strategies of both firms under Stackelberg competition. Exploiting its strategic advantage as a leader, the low cost firm will enter the market later, exit sooner and install more capacity than in the Cournot case because the preemptive threat of firm B is lower. However, it will still choose a lower demand parameter for entry and for exit, along with a lower production capacity than in the monopolistic setting. While the option to lead adds value to the low cost firm, it is worth more than as a player in Cournot duopoly but worth less than as a monopolist. The Stackelberg competition can therefore be seen as an intermediate situation between Cournot competition and monopoly. As the strategic advantage of the low cost firm gets arbitrarily large ($c_B \gg c_A$) the situation for the Stackelberg leader approaches the monopolistic case. Because the high cost firm B does not have an equal role with firm A in Stackelberg competition, its competitive disadvantage becomes more pronounced than under Cournot competition. This leads to even higher entry and exit demand levels as well as higher optimal production quantities. Hence, its actions deviate more from the monopolistic case resulting in a lower firm value of the follower. These basic results that should hold for reasonable parameter values are summarized in Table 5.3.

Of course, the entry, exit and capacity choice model under perfect informa-tion can be extended in several ways in order to value more complex competitive interaction effects. It would not be very difficult to identify other stochastic pro-cesses for the demand function. For example the demand parameter could also be modelled as a mean-reverting process or in a regime switching model to take account of life cycles that frequently occur in product markets.[22] Furthermore, other real options can be integrated as well in order to make the model more realistic. A natural extension would be to allow for production flexibility as a response to competitive entry and exit. In addition to the perfect information that was assumed in the example, asymmetric information between the players can also be examined. However game theoretic analysis is highly sensitive with respect to its input parameters. The resulting equilibrium very much depends on a number of questions. How many competitors are in the market? What is the order of the strategic moves of the competitors? Are there information asymmetries between market participants? Do players cooperate? Which particular form take the cost and demand functions? How robust is the resulting equilibrium to changes in the model specification? For the example considered here the answers to these questions were assumed to be known. However, for practical situations some of the above questions may be hard to answer precisely enough in order to determine the resulting equilibrium.

To conclude, we have seen that game theory can be exploited to value and explain competitive interaction effects in real option pricing. It therefore represents an important extension of the real option approach in a competitive environment without the naive assumptions of either exclusiveness of real options or completely exogenous competition. In terms of the stochastic control framework, the example of the joint entry, exit and capacity choice decision in duopoly was solved as an impulse control game. Although the complexity of the model increased significantly, the insights gained may justify the use of game theory for real option valuation purposes in research and practice. While game theoretic ideas are not new to the real options literature, a thorough treatment of the subject is still missing. In showing how game theory can be incorporated into the stochastic control framework of real options a first step in that direction has been made. However, from a research point of view option games are still in their infancy and much empirical and theoretical work is to be done in the future.

5.2   Time Delay Effects

Another important limitation to the financial option analogy of real options is the frequently observable occurrence of *time lags* in real life investments between the date of the exercise decision and the time when it finally takes effect.[23] The resulting *investment lag* can be caused for instance by the time it takes to build the project (construction time of a production facility, product development period etc.) or the time it takes to complete a disinvestment (time to find a buyer, contractual constraints, negotiations with unions etc.). More generally, an investment lag is the time between the decision to alter the state of the firm and the time when it enters the new state. Although most models in the real option literature assume that the investment decision is implemented immediately, the consideration of time

---

[22]See Chapter 6.
[23]See Section 2.2.2.

delay effects can dramatically change the investment uncertainty relationship. It is therefore important to accurately take account of investment lags in pricing real options and in determining the optimal exercise strategies. By making use of a model of investment lags introduced by Bar-Ilan/Strange [12], the aim of this section is again to develop a graphical representation for time lags, to show how the exemplary model of Bar-Ilan and Strange fit into the stochastic control framework, and to interpret the importance of recognizing investment lags for valuation and strategy choice.

The model considered by Bar-Ilan and Strange consists of a firm that faces uncertainty about its output price P which is exogenously given and follows the stochastic process

$$dP_t = (r - \delta) P_t \, dt + \sigma P_t \, d\tilde{B}_t, \qquad P_0 = p \tag{5.75}$$

under the equivalent martingale measure $\tilde{\mathbb{P}}$. Assume that the inactive firm has the opportunity to invest in a project to obtain a technology producing forever one unit of output per unit of time, by paying a fixed investment cost which is completely sunk. Since construction will consume a certain prespecified time h, the firm will not receive revenues from the project until h periods after the decision to invest is made. At the end of the construction period the firm can start producing by paying a marginal constant cost of production c and earns the uncertain price P while it possesses the possibility to abandon operation. If the firm is inactive it waits for good news that makes investment favorable. The value of the inactive firm consists solely of its option to wait to invest, which is denoted by $V_0(P)$. If the firm is operating, its value $V_1(P)$ consists of its running profit and the option to later abandon the project. In order to enter or exit the market an investment outlay of $I > 0$ and a disinvestment outlay of $E \geq 0$ has to be made.[24]. So far, the described model is the same as for the entry and exit model of Dixit/Pindyck [72]. However, once the firm has decided to invest it does not immediately start producing but has to enter a construction phase which lasts h time periods before completion. The construction phase can be seen as an auxiliary state during which the firm has the value $V_C(s, P)$ where $0 \leq s \leq h$ is the remaining time to completion of the construction phase. For simplicity, it is assumed that there are no cash in or outflows during the construction phase. Furthermore, the lump investment I is payable at completion time but is considered sunk by the time the investment decision is made. It is therefore never optimal to abandon the project during the construction phase. Lastly, the model assumes an infinite time horizon.

In order to find a graphical representation of this model first notice that we only have to consider the two costly reversible states inactive and operating, since once the investment decision has been made the earliest possible time of abandonment is right after completion of the construction phase. However, the firm value in the auxiliary state construction is crucial for optimization purposes and is for

---

[24]Once having abandoned the project, the firm is again holding the option to invest with value V_0. Thus, the switches between the states are costly reversible.

that reason also displayed in the graphical representation. The description of the states involved is an easy exercise of extracting the required information from the above paragraph. The switching cost function for the states inactive ($Z_t = 0$) and operating ($Z_t = 1$) is given by the matrix

$$H(Z_t, 1 - Z_t) = \begin{pmatrix} 0 & Ie^{-r\tau} \\ E & 0 \end{pmatrix} \tag{5.76}$$

where the switching costs are independent of time t and output price P. The instantaneous cash flow earned by the operating firm is

$$f_1(P) = P - c \tag{5.77}$$

while it is zero if the firm is inactive or in construction. Since the states inactive and operating are costly reversible with the transient construction phase the modelling framework is a generalized switching option.

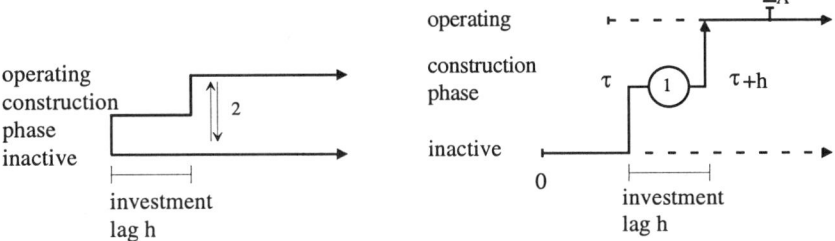

Figure 5.13: Modelling element time lag as generalized switching option

Figure 5.14: Modelling element time lag as generalized timing option

In principle, there exist two different ways to represent time lags in the graphical decomposition framework. The first is in terms of a generalized switching option as displayed in Figure 5.13. The kink of length h stands for the time lag when switching from inactive to operating is triggered. However, the discussed model can as well be represented in explicit form using the modelling elements of irreversible switches and abandonment as can be seen in Figure 5.14. The second is more intuitive and provides a more detailed description of the decision situation but does not allow for reentry after having abandoned the project. Therefore, the modelling element time lag of a generalized switching option in Figure 5.13 is an infinite sequence of modelling elements of the generalized timing option. This indicates again that an impulse control model is nothing but a sequence of stopping problems.

In order to solve for the value of the generalized switching option with time lag between state inactive and operating, we formulate the corresponding impulse control problem. The optimization problem is given similar to the entry and exit decision of the monopolistic firm of the above section as

$$V_0(P) = \sup_{w \in \mathcal{W}} \tilde{\mathbb{E}} \left[\int_0^\infty e{-rt} \, f_{Z_t}(P_t) \, dt - \sum_{i=1}^\infty e^{-r\theta_i} H(Z_t, 1 - Z_t) \right] \quad (5.78)$$

where it is assumed that the firm starts in state inactive (0). An admissible impulse control strategy w now consists of a sequence of stopping times θ_i and control actions $Z_{\theta_i} = \zeta_i \in \{0, 1\}$ according to

$$w = (\theta_1, \theta_2, \ldots, \theta_{2i-1}, \theta_{2i}, \theta_{2i+1}, \ldots; 1, 0, \ldots, 1, 0, 1, \ldots) \quad \in \mathcal{W}. \quad (5.79)$$

It is clear that when starting in state inactive the first possible action is to switch to the operating state $\zeta_1 = 1$ and the second back to the state inactive $\zeta_2 = 0$ and so forth. Therefore all odd impulse controls ζ_{2i-1} are 1 and all even impulse controls ζ_{2i} are 0 ($i = 1, 2, \ldots$). Furthermore because the firm completes the construction phase in any case and does not abandon in midstream, the earliest point in time when it abandons is at least h time periods after the decision to enter the construction phase has been made. Consequently, the set of all admissible impulse control strategies $\mathcal{W}$ consists of

$$\mathcal{W} = \{\{\theta_i, \zeta_i\}_{i \in \mathbb{N}} : \theta_{2i-1} + h < \theta_{2i} < \theta_{2i+1}, \ \zeta_{2i-1} = 1, \ \zeta_{2i} = 0\}. \quad (5.80)$$

Thus, the problem is to find the solution of the impulse control problem with the constraint that changing the state of the system from 0 to 1 consumes a time period of h. Although the construction phase does not explicitly enter the optimization problem, we need to know the value of the firm in construction V_C in order to solve the impulse control problem. This becomes obvious in the formulation of the corresponding sufficient quasi-variational inequalities. Starting with switching from inactive to operating, the maximum operator in the present case is

$$\mathcal{M}V_0(P) = V_C(h, P) - e^{-rh} I. \quad (5.81)$$

That means, when the firm decides to enter the construction phase which it has to complete, it exchanges its value as an inactive firm with the value of a firm under construction, net of the present value of the investment outlay payable at the end of construction. V_C in turn depends on the outcome of the price after the construction phase is finished. If the price will go up during the time to completion, the firm will start operation and will therefore be worth V_1. If on the other hand the output price is sufficiently low after the construction phase, the firm will immediately abandon. In this case the value of the firm directly after completion will be the value of the inactive firm minus exit cost E. This gives rise to the second maximum operator for switching from state operating to inactive

$$\mathcal{M}V_1(P) = V_0(P) - E. \quad (5.82)$$

Thus, once having entered the construction phase with s time periods to go before completion, the value of the firm is

$$V_C(s, P) = \tilde{\mathbb{E}}\left[e^{-rs} \max\{V_1(P(s)), V_0(P(s)) - E\}\right] \qquad (5.83)$$

where $P(s)$ is the price s time periods ahead.

Hence, the set of quasi-variational inequalities for switching from inactive to operating is given by

$$\frac{1}{2}\sigma^2 P^2 \frac{d^2 V_0(P)}{dP^2} + (r - \delta)P \frac{dV_0(P)}{dP} - rV_0(P) \leq 0 \qquad (5.84)$$

$$V_0(P) \geq V_C(h, P) - e^{-rh} I,$$

One of the inequalities is an equality.

and for switching from operating to inactive by

$$\frac{1}{2}\sigma^2 P^2 \frac{d^2 V_1(P)}{dP^2} + (r - \delta)P \frac{dV_1(P)}{dP} - rV_1(P) + (P - c) \leq 0, \qquad (5.85)$$

$$V_1(P) \geq V_0(P) - E.$$

One of the inequalities is an equality.

Solving the resulting ODEs of (5.84) and (5.85) that V_0 and V_1 have to satisfy in the continuation region yields the familiar functions

$$V_0(P) = A_1 P^{\beta_1} \qquad (5.86)$$

and

$$V_1(P) = A_2 P^{\beta_2} + \frac{P}{\delta} - \frac{c}{r} \qquad (5.87)$$

where β_1 and β_2 are again the positive and negative roots of the characteristic equation (4.86)

$$\beta_{1/2} = \frac{1}{2} - \frac{r - \delta}{\sigma^2} \pm \sqrt{\left(\frac{r - \delta}{\sigma^2} - \frac{1}{2}\right)^2 + \frac{2r}{\sigma^2}}. \qquad (5.88)$$

Once again we have used the side conditions that as P gets small V_0 approaches zero, which allows us to eliminate the negative root part of the general solution of V_0. Vice versa, the positive root part of V_1 was eliminated because V_1 must approach the static net present value of the operating firm ($P/\delta - c/r$) as P gets arbitrarily large. As above the value of the inactive firm V_0 consists exclusively of the option to wait. The first summand on the right side of V_1 reflects the option value to abandon while the rest represents the expected net present value of the future cash flows of the operating firm.

Next we determine the value of the firm under construction $V_C(s, P)$. Since at the end of the construction phase the firm is either operating or abandons immediately, there exists an abandonment price P_l for which

$$V_1(P_l) = V_0(P_l) - E. \tag{5.89}$$

P_l is the abandonment trigger price of the operating firm. Thus, $V_c(s, P)$ can be found by

$$
\begin{aligned}
V_c(s, P) &= \tilde{\mathbb{E}}\left[e^{-rs} \max\{V_1(P(s)), V_0(P(s)) - E\} \right] \qquad (5.90) \\
&= \tilde{\mathbb{E}}\left[e^{-rs} V_1(P(s)) \mid P(s) > P_l \right] \\
&\quad + \tilde{\mathbb{E}}\left[e^{-rs} (V_0(P(s)) - E) \mid P(s) \le P_l \right] \\
&= \int_{P_l}^{\infty} e^{-rs} \left(A_2\, P^{\beta_2}(s) + \frac{P(s)}{\delta} - \frac{c}{r} \right) \varphi_{log}(P(s))\, dP(s) \\
&\quad + \int_0^{P_l} e^{-rs} \left(A_1\, P^{\beta_1}(s) - E \right) \varphi_{log}(P(s))\, dP(s)
\end{aligned}
$$

where $\varphi_{log}(P(s))$ is the density function of the output price s time periods in the future. Since P follows geometric Brownian motion, $P(s)$ is lognormally distributed with mean $\mu = \log P + (r - \delta - \frac{1}{2}\sigma^2)s$ and variance $v^2 = \sigma^2 s$. The kth moment about the origin of a lognormal random variable $y = \log x \sim N(\mu, v^2)$ is given by

$$\mathbb{E}\left[x^k \right] = e^{k\mu + \frac{1}{2}k^2 v^2} \tag{5.91}$$

while the higher partial moment on the interval $[x_0, \infty)$ can be obtained as

$$\mathbb{E}\left[x^k \mid x \ge x_0 \right] = (1 - \Phi(u - k\sigma))\, e^{k\mu + \frac{1}{2}k^2 v^2}. \tag{5.92}$$

Here, u stands for

$$u = \frac{\log x_0 - \mu}{v} \tag{5.93}$$

and $\Phi(.)$ denotes the cumulative standard normal distribution function. Hence, we have

$$
\begin{aligned}
\int_{P_l}^{\infty} P^k(s)\, \varphi_{log}(P(s))\, dP(s) &= (1 - \Phi(u - k\sigma))\, \tilde{\mathbb{E}}\left[P^k(s) \right], \quad (5.94) \\
\int_0^{P_l} P^k(s)\, \varphi_{log}(P(s))\, dP(s) &= \Phi(u - k\sigma)\, \tilde{\mathbb{E}}\left[P^k(s) \right].
\end{aligned}
$$

Substituting back into (5.90) yields

$$V_c(s, P) \;=\; e^{-rs}\Bigg(A_2\Big(1 - \Phi(u - \beta_2\sigma)\Big)\tilde{\mathbb{E}}\big[P^{\beta_2}(s)\big] \tag{5.95}$$

$$+ \Big(1 - \Phi(u - \sigma)\Big)\frac{\tilde{\mathbb{E}}\,[P(s)]}{\delta}$$

$$- \Big(1 - \Phi(u)\Big)\frac{c}{r} + A_1\,\Phi(u - \beta_1\sigma)\tilde{\mathbb{E}}\big[P^{\beta_1}(s)\big] - \Phi(u)\,E\Bigg).$$

After some calculations using the definition of the characteristic roots β_1 and β_2 as well as the relationship of the moments in (5.91), we finally get

$$V_c(s, P) \;=\; A_2\Big(1 - \Phi(u - \beta_2\sigma)\Big)P^{\beta_2} + \Big(1 - \Phi(u - \sigma)\Big)\frac{P\,e^{-\delta s}}{\delta} \tag{5.96}$$

$$+ \Big(1 - \Phi(u)\Big)\frac{c\,e^{-rs}}{r} + A_1\Phi(u - \beta_1\sigma)\,P^{\beta_1} - \Phi(u)E\,e^{-rs}$$

where u is defined as

$$u = \frac{\log(\frac{P_l}{P}) - (r - \delta - \frac{1}{2}\sigma^2)s}{\sigma\sqrt{s}}. \tag{5.97}$$

Turning now to the shape of the continuation regions of the states. As for the entry and exit decision without time lags, the firm will invest in the venture as soon as the price rises high enough and reaches a trigger price P_h. At the boundary of the continuation region ∂C, which already belongs to the intervention region, the second inequality of (5.84) must be an equality which yields the value matching condition for the investment trigger P_h,

$$V_0(P_h) = V_c(h, P_h) - e^{-rh}I. \tag{5.98}$$

Since by construction the value function V is stochastically C^2 it is at least once continuously differentiable at the boundary ∂C. This results in the smooth pasting condition

$$V_0'(P_h) = \frac{\partial V_c(h, P_h)}{\partial P}. \tag{5.99}$$

The same arguments lead to the value matching and smooth pasting condition for the abandonment trigger P_l:

$$V_1(P_l) \;=\; V_0(P_l) - E, \tag{5.100}$$

$$V_1'(P_l) \;=\; V_0'(P_l). \tag{5.101}$$

Equations (5.98) to (5.101) form again a system of nonlinear equations that can be solved numerically for the two unknown constants A_1 and A_2 as well as for the entry and exit triggers P_h and P_l.

Bar-Ilan/Strange examined the properties of the solution with the base case parameters as follows. The annual riskless interest rate is $r = 0.025$. The annual rate of return shortfall δ for the price process in risk neutral equilibrium is 0.025. The annual volatility σ varies from 0 to 25%. For normalization the marginal production cost c is set to a value of 1. The fixed investment cost is assumed to be $I = 1$ which corresponds to a ratio of 40 to 1 between the marginal production cost c and the annualized cost of capital rI. The considered exit cost E is either 0 or 1. Figure 5.15[25] shows the effect of changes in uncertainty σ^2 for the case of no investment lag ($h = 0$) and an investment lag of $h = 6$ years when exit costs E are 0 on the optimal investment triggers P_l and P_h. In the deterministic case ($\sigma^2 = 0$) the investment lag h does not have an effect on the optimal entry and exit strategy. P_l is equal to 1 while P_h is equal to 1.025, taking account of the annualized cost of capital $rI = 0.025$. As in the classical entry and exit model of Dixit/Pindyck [72] with no investment lag, higher uncertainty raises the entry trigger P_h and lowers the exit trigger P_l, respectively. As usual more uncertainty delays both investment and abandonment leading to a larger hysteresis effect.

However, in the presence of investment lags the impact of increasing uncertainty is surprising. While for an investment lag of $h = 6$ the effect of uncertainty on the exit threshold P_l is fairly standard[26] the entry trigger price P_h dramatically changes. For low levels of uncertainty P_h rises to a local maximum before it drops, even under the certainty trigger price, and rises again afterwards. For all values of σ^2 the entry threshold with time lag is below the entry threshold without time lag. Thus, there are three important results. The first is that investment lags lead to earlier investment than in the no lag case. Second, that investment in the presence of lags might occur earlier under uncertainty than under certainty. And third, that more uncertainty might even hasten investment for some parameter constellations.

The intuition behind these striking results needs to be explained. In the case of no investment lags it is found in the literature that more uncertainty delays investment, since an increase in uncertainty raises the benefit of waiting but leaves the opportunity cost of a further delay unaffected. This is because the opportunity cost of waiting consists of the instantaneously foregone profit by not investing, while on the other hand higher uncertainty causes a higher downside risk for the firm. Since the firm can enter immediately, higher uncertainty keeps the opportunity cost unchanged. Consequently, higher uncertainty delays investment.

In contrast, the effect of the investment lag that leads to earlier investment is caused by increasing opportunity cost of waiting in the presence of the option to abandon. A firm that has to enter a construction phase first before starting operations has not to consider the foregone profit during the waiting period but the foregone profit during an equally long time period after completion of the construction phase. Because the firm has the option to abandon, these expected foregone future profits

---

[25]See Bar-Ilan/Strange [12], table 1, p. 616.

[26]This is as expected since there is no time lag when exiting the market. However, the exit trigger P_l is lower with lags than without because of the time it takes to reenter the market.

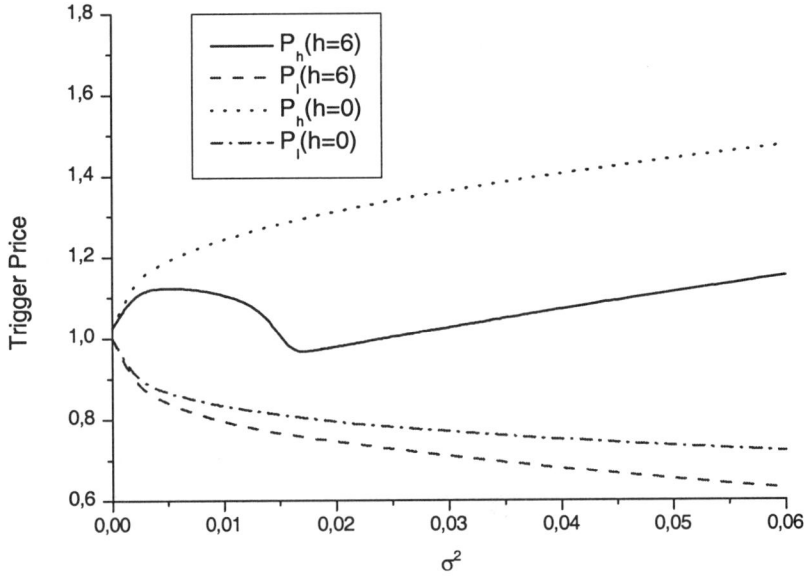

Figure 5.15: Impact of increasing volatility on entry and exit triggers with and without time lag, $E = 0$

are truncated from below leading to higher opportunity cost of waiting. Therefore, the opportunity cost of waiting also rises with increasing uncertainty. Depending on the specific situation the increasing opportunity cost of waiting might dominate the increasing benefit of waiting, leading to earlier investment. In Figure 5.15 this effect can be observed for σ^2-values between 0.006 and 0.018. The ability to abandon is crucial for the reverse effect of time lags. If abandonment is more costly, the protection of future expected profits against low output prices is less pronounced and will tend to converge to the classical results. For the same parameter values as above the results for costly exit ($E = 1$) are presented in Figure 5.16.[27] As can be seen from the figure the reverse effect that more uncertainty hastens investment occurs only for values of σ^2 larger than 0.028 and is not as emphasized as above. Although $E = 1$ is already an extreme case because it implies that disinvestment is as expensive as investment ($I = 1$), a further rise in E would cause the model to collapse to the classical case. If for example we have $rE > c$, then abandonment will never be optimal because the annualized exit cost will be larger than the

---

[27] See Bar-Ilan/Strange [12], table 2, p. 617.

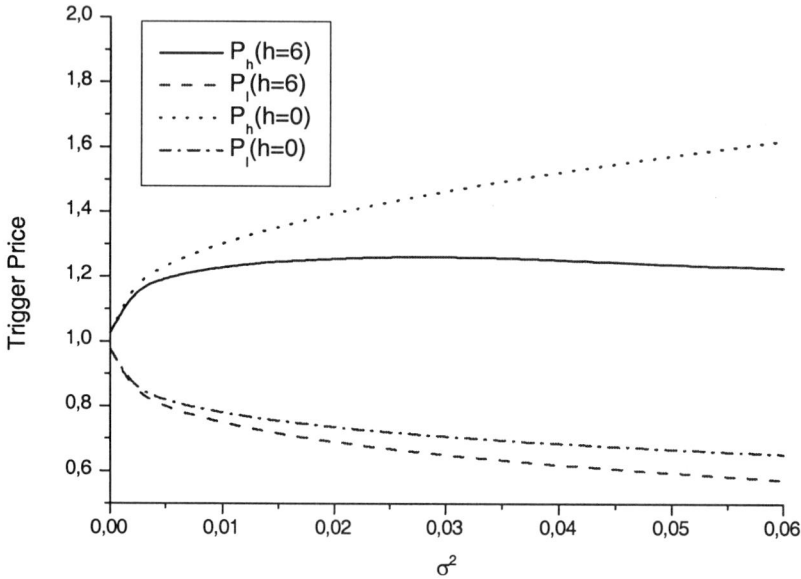

Figure 5.16: Impact of increasing volatility on entry and exit triggers with and without time lag, $E = 1$

variable marginal production cost. In this case, Bar-Ilan/Strange showed that the investment trigger P_h is given by

$$P_h = \frac{\beta_1}{\beta_1 - 1} \frac{\delta}{r} \underbrace{e^{-(r-\delta)h}}_{(*)} (c + rI) \qquad (5.102)$$

which is exactly the trigger price of the option to invest with fixed capacity in equation 4.100, corrected by the expected growth of P during the construction phase $(*)$. For this trigger price the same comparative statics results can be obtained as in the classical model. However, in realistic scenarios where the exit cost E is less than the investment outlay I, uncertainty still hastens investment. Thus considering investment lags in determining optimal investment decisions cannot be neglected.

Finally, in Figure 5.17[28] the effect of different investment lags on the entry and exit trigger prices is examined for $\sigma^2 = 0.01$ and $E = 1$. As can be observed from the figure, P_l and P_h first decrease in h before they finally increase. This

---

[28] See Bar-Ilan/Strange [12], table 3, p. 618.

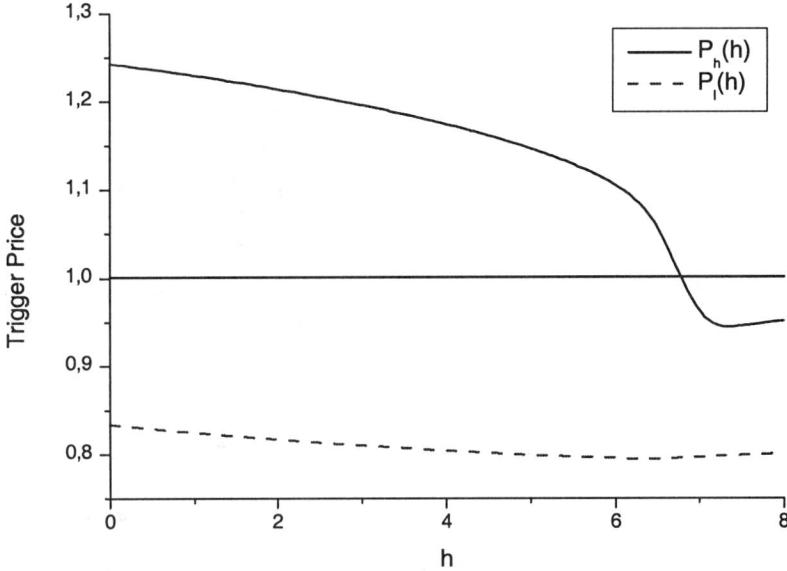

Figure 5.17: Impact of increasing time lag h on trigger values, $E = 1$ and $\sigma^2 = 0.01$

effect is more pronounced for P_h than for P_l. The reaction of P_h on the value of h is achieved by two opposite effects. On one hand a larger h raises the variance of possible market prices at the end of the construction phase which leads to an increase in the option to wait. This effect was also observed in a different model of time to build by Majd/Pindyck [150]. On the other hand because of the truncated downward risk due to the abandonment option the opportunity cost of waiting increases with the time lag as well. This latter effect dominates the first one up to $h = 7$ and results in decreasing entry trigger prices P_h until the first effect takes over for values of h above seven years.

The consideration of investment lags has some severe practical implications. As Bar-Ilan/Strange point out, the presence of investment lags frequently occurring in the chemical industry, the electric utility industry, or real estate can, together with other effects caused for example by competition, explain overbuilding phenomena or investment in industries with chronic excess capacity. In all of these industries significant time lags lead to early investment despite the currently unfavorable market conditions because firms fear the threat of not being in the market when the industry eventually recovers.

Another important implication of time delay effects was discussed in a more recent article by Bar-Ilan/Strange [14]. They studied an investment project with two sequential construction phases, each with a certain time lag. They showed that for a similar model as the one discussed here the investment trigger for the first construction stage may be below the investment trigger for the second construction stage. This is a surprising result because for the first investment decision the whole investment sum has to be taken into account, while for the second investment decision only the (lower) investment cost of the second stage has to be considered. Orthodox theory would suggest that the first trigger is always higher than the second one. In fact, this holds true if the firm does not have the option to suspend after the completion of the first construction phase. If however the firm can wait before starting the second construction phase without having to pay high suspension cost, it may be optimal to invest earlier into the first stage than into the second stage. Important for a lower first investment trigger is how the investment costs and time lags are portioned among the two construction stages. Especially investment projects with relatively small investment cost in the first and relatively short time lag in the second construction phase admit the above mentioned reverse behavior. These findings are consistent with the one time lag case where longer investment lags and lower investment cost also lowered the investment trigger. An increase in uncertainty leads to the same effect because the first trigger rises less than the second trigger.

However, notice that this effect is not caused by a learning option where only the active firm experiences uncertainty resolvence. The reason that the first investment trigger might be lower in a two stage construction phase with time lags is that a firm which has already completed the first stage and is waiting to enter the second stage can benefit faster from improved market conditions than otherwise inactive firms which are yet to complete both construction stages.

Chapter 6

Case Study: Flexibility in the Manufacturing Industry

So far the analysis has concentrated on the introduction of the stochastic control framework and the graphical decomposition method. Afterwards we extended the framework to deal with several special problems that occur with real options. But we have still not fully exploited the potential of the stochastic control framework to handle complex option interaction models. For that reason a realistic real option interaction model of production flexibility that might occur in practice will be presented in this chapter along with its main implications from an economic point of view as well as from a valuation technical point of view.

In order to make the investment decision problem more realistic some assumption are weakened that have been made in the above chapters. First of all, we drop the assumption of infinite time horizons. In considering a finite time horizon, the American type options nested in the generalized timing and switching option have to be valued numerically. Especially, it will be demonstrated how numerical solutions of the associated general impulse control problems with different control variables can be found. A subject that has been rarely addressed in either the literature on impulse control or the real option literature.[1] The analysis below therefore represents an important step in closing the gap between the theoretical work on impulse control and its use in practical applications. Furthermore, we will allow for

---

[1]Apart from perpetual entry and exit models, which are the easiest conceivable impulse control problems in real option pricing because they consist of only two states between which switching takes place, there exists only one article by Mauer/Triantis [157] dealing with impulse control. They numerically find a solution to an impulse control problem, however, without explicitly noting it. Korn [121] discusses an analytic approximation for a special simplified portfolio optimization problem. Liu/Yao [146] present a solution in terms of a series expansion to a special time-homogeneous impulse control problem in re-engineering. Beyond that all other articles on impulse control either focus on theoretical topics or find analytic solutions of specific simple examples using barrier controls which are different from the indirect control case in real option pricing.

more realistic stochastic processes than the above used geometric Brownian motion. In using a regime switching model of different geometric Brownian motions, introduced by Bollen [28], we are able to model demand as a stochastic product life cycle. The regime switching model implicitly consists of both market priced and private risk. Thus, this is another example of how private risk can be incorporated into the stochastic control framework.

Another goal of this chapter is to demonstrate that the graphical decomposition of the contingency structure of interacting real options is a very powerful tool in structuring complex real option problems and easily transforming them into numerical solution techniques using a standardized procedure. In addition, this is achieved without losing the general overview and the good communication properties a graphical representation admits in contrast to a pure mathematical representation. Furthermore, the model shows how to work through the steps of the real option valuation process; especially in finding the optimal strategy, the value of the investment and its risk profile.

6.1 Real Options and Volume Flexibility

Consider a company that expects to soon finish development of a new product line. The new product line is intended to replace an old one whose demand has sincerely dropped recently. Unfortunately, due to new technical features the new product line cannot be produced using the old products´ production facilities. For that reason, management has to decide over building a new plant. The main source of uncertainty the company faces is uncertainty over future demand of the new product. From industry analysis it is known that approximately each seven to eight years a new product line is introduced in the market that substitutes the old one. Thus, with an overlapping time of about two years between bringing the new product successfully to the market and the complete erosion of demand of the old product line the average life of a product line, is approximately ten years. The production technology of the new plant allows for flexible manufacturing in both the production rates and capacity. With varying production rates the firm seeks to compensate short time fluctuations in demand while significant upside or downside changes in demand may be opposed by expanding or contracting the overall capacity of the production facility, which is assumed to be possible in between a certain range of capacity levels. Of course, management has the option to wait to invest and choose its entry capacity. This kind of investment problem occurs frequently in all kinds of industries (e.g., car manufacturing, chemical and pharmaceutical industry, chip industry etc.) and the basic question to answer is, of course, if the investment in the new production facility is favorable or not. If we want to accurately value the project in order to make the investment decision, it is obvious that a standard net present value analysis will be grossly misleading since it does not take account of the incorporated flexibilities to adjust production rates, capacity levels, optimal timing and intensity of investment etc., but rather values a predetermined scenario with an ad hoc determined strategy. A real option analysis, however, seems to be the right choice. Yet, due to the infinite number of options to contract, expand, costly suspend, wait and abandon embedded into the above described decision situation, performing a real option valuation using the generic real options mentioned in the literature is rather complicated if not impossible. As it will turn out the above decision situation can be easily displayed, modelled and valued in the stochastic control framework using combinations of generalized timing and switching options.

In an age of uncertainty and environmental discontinuities, manufacturing flexibility in the described way has become a crucially important topic for production oriented companies. It is ranked as a strategic priority together with, e.g., cost and quality. The importance of the real option approach in manufacturing flexibility is in its ability to relate uncertainty to the value of flexibility. It therefore represents a measure for production flexibility. According to Sethi/Sethi [200] production flexibility can be broadly distinguished into process flexibility and volume flexibility. Process flexibility refers to the set of parts that can be produced without a major set-up. It is sometimes called product-mix flexibility in the literature. Volume flexibility in turn defines the ability to operate profitably at different output levels including different production rates and the ability to adjust capacity. The model developed here is mainly concerned with the value of volume flexibility.

In the real option literature production flexibility[2] is often referred to as operating options consistent with the real option classification of Section 2.2.1. Several articles focussing on operating options in production have been published.[3] So far, the literature on volume flexibility is restricted to either dealing with incremental rather than lumpy investment[4] or considering only one single expansion or contraction decision.[5] In each of these articles certain options were isolated in order to allow either for analytic solutions or for the use of simple generic real options. However, none of the discussed models allowed for costly reversible switches leading to impulse control problems. Furthermore, the source of uncertainty of the different models was usually assumed to follow geometric Brownian motion. Especially, in the case of demand uncertainty which is normally assumed in real option models of production flexibility the assumption of demand (or a demand function parameter as in Section 5.1.2) following a geometric Brownian motion is unrealistic. As pointed out by Bollen [28] demand for goods in many industries can be better characterized by product life cycles. Amongst others, product life cycles

---

[2]A non-option approach to production flexibility using dynamic programming can be found in Metters [168], Harrison/Van Mieghem [91] and Van Mieghem [228] and the references therein.

[3]See Triantis/Hodder [219], Kulatilaka [123], Kulatilaka/Marks [129], Kulatilaka [124] and Chen/Kensinger/Conover [47] for process or product-mix flexibility with fixed capacity and two products; Kamrad/Ernst [107] for the formulation of a model with $m > 2$ products with input price and output yield uncertainty; Kogut/Kulatilaka [118], Mello/Parsons/Triantis [164] and Huchzermeier/Cohen [100] for the process and product-mix flexibility of multinational firms under exchange rate risk.

[4]See Cortazar/Schwartz [63], Cortazar/Schwartz/Löwener [62], He/Pindyck [93] and Kamrad/Lele [108].

[5]See Cortazar/Schwartz/Salinas [64].

of demand can be caused by market saturation, by the introduction of competing products or substitutes, the development of a superior technology, or just changing consumer tastes. Depending on the model, several different phases for a product life cycle can be identified. The model of Bollen [28] that is adopted here uses two phases which consist of a growth and a decay regime that are separated by a random switching time. In the growth regime, demand follows an upward sloping geometric Brownian motion, while in the decay regime the demand follows a downward sloping geometric Brownian motion combining the mathematical tractability of a GBM-model with a more realistic stochastic process for demand dynamics.

The sections of this chapter are organized as follows. In Section 6.2 the real option model of production flexibility will be described in mathematical terms together with its representation in the graphical decomposition method. Section 6.3 presents how the model can be solved numerically using finite differences. The numerical valuation results and the optimal operating policies are discussed in Section 6.4. The optimal policies are used to simulate the option interaction model in Section 6.5.

6.2   Model

In order to specify the above mentioned decision situation further, note that the flexibility contained in the project is a combination of strategic and operating options. The firm has the strategic option to choose the time when the new plant is built and the size of the plant if it is built. Therefore this is a timing and intensity option which creates a new strategic asset. The options to adjust production rates and capacity are operating options according to the definition of Section 2.2.1. In having the means to adjust capacity in a certain range of values through the life of the project, the firm holds infinitely many options to expand and contract the scale of the project. Moreover, in controlling production rates the firm has the option to costly suspend operations whenever it is optimal to do so.

The source of uncertainty that drives management´s decisions is uncertainty in demand. Similar to Section 5.1.2 the demand function is assumed to follow the relationship

$$Q_t \;=\; \Theta_t \,-\, \lambda\, P_t \tag{6.1}$$

where Q_t represents the demand for products at time t, Θ_t is a stochastic demand parameter, $\lambda > 0$ stands for the constant slope of the demand function and the price P_t is implicitly determined by the assumption that in market equilibrium there exists an instantaneous clearing of demand and supply. In the product life cycle model considered here, Θ_t is assumed to start in a growth regime g before it switches at a random time τ to a decay regime d. In the growth regime ($t \in [0, \tau)$), Θ_t follows a geometric Brownian motion with positive drift parameter μ_g and constant positive volatility σ according to

$$d\Theta_t = \mu_g\, \Theta_t\, dt \,+\, \sigma\, \Theta_t\, d\tilde{B}_t, \qquad \Theta_0 \text{ given and positive}, \tag{6.2}$$

while in the decay regime ($t \in [\tau, T]$) the overall demand is expected to decrease and follows a geometric Brownian motion with negative drift parameter $\mu_d < 0$:[6]

$$d\Theta_t = \mu_d \, \Theta_t \, dt + \sigma \, \Theta_t \, d\tilde{B}_t.$$ (6.3)

This implies that once having reached the decay regime the demand parameter cannot switch back to the growth regime. Of course, in more detailed regime switching models more life cycle phases can be included. The random switching time τ is assumed to be (truncated) normally distributed with mean μ_τ and standard deviation σ_τ on the interval $[0, T]$. Although the specification of the switching time distribution is critical for the value of the project and the optimal operating strategy, assuming a normal distribution is a first reasonable approximation of reality. However, other distributional assumptions can be easily incorporated into the analysis. The resulting regime switching probabilities can be determined using Bayes´ rule. For example, being in the growth regime at time t the probability of switching to the decay regime during the next small time interval dt is given by

$$P(\tau \leq t + dt | \tau > t) \quad = \quad \frac{P(\tau \in (t, t + dt])}{P(\tau > t)} = \frac{P(\tau \in (t, t + dt])}{1 - P(\tau \leq t)}$$

$$= \quad \frac{\varphi\left(\frac{t - \mu_\tau}{\sigma_\tau}\right)}{1 - \Phi\left(\frac{t - \mu_\tau}{\sigma_\tau}\right)} dt = l(t) \, dt$$ (6.4)

where $\varphi(x)$ and $\Phi(x)$ are the (truncated) standard normal density and distribution functions on the interval $[0, T]$, respectively. $l(t)$ is the *switching rate* from growth to decay at time t. Similarly, the transition probability of staying in the growth regime during the time interval $(t, t + dt]$ can be obtained by

$$P(\tau > t + dt | \tau > t) \quad = \quad 1 - P(\tau \leq t + dt | \tau > t)$$ (6.5)

$$= \quad 1 - \frac{\varphi\left(\frac{t - \mu_\tau}{\sigma_\tau}\right)}{1 - \Phi\left(\frac{t - \mu_\tau}{\sigma_\tau}\right)} dt = 1 - l(t) \, dt.$$

Once having reached the decay regime the probability of switching is zero by assumption and, consequently, the probability of staying in the decay regime is one.

For the regime switching model to work properly it has to be assumed that managers as well as capital markets know the current regime. This might be cumbersome because the demand parameter might randomly increase although it is already in the decay regime leading to a wrong assessment by management. However, in many industries the beginning of the decay regime is relatively clear cut because of announcements of new technologies or the introduction of substituting products.[7]

---

[6]See Bollen [28], p. 2.
[7]See Bollen [28], p. 10.

In mathematical terms, the resulting filtration in the growth regime $\mathcal{F}_t^g$ consists solely of the information set $\mathcal{F}_t$ generated by the Brownian motion $\{\tilde{B}_t\}_{0 \leq t < \tau}$, i.e., $\mathcal{F}_t^g = \mathcal{F}_t$ because τ is not yet revealed. However, in the decay regime the filtration additionally contains the information about the past regime switching time τ, i.e., $\mathcal{F}_t^d = \mathcal{F}_t \cup \{\tau = t_1\}$. Thus, at t_1 switching is revealed.

Lastly, we turn to the question of finding an equivalent martingale measure $\tilde{\mathbb{P}}$ for the regime switching model. Although it may be a difficult task to find a twin asset for the stochastic nature of the demand parameter, it is assumed here that the risk of the product life cycle with a priori fixed τ, i.e., $\sigma_\tau = 0$, is priced by the market and that μ_g and μ_d already represent the risk neutral growth and decay rates in the risk neutral world subject to the measure $\tilde{\mathbb{P}}_B$ with the new Brownian motion $\tilde{B}_t$. However, the risk introduced by the random switching time τ is assumed to be private to the firm with the probability measure $\mathbb{P}_\tau$ known from the switching probabilities. As already discussed there are several approaches to incorporating private risk into a real option analysis. Although individual investors may have different utility functions towards the regime switching risk, again risk neutrality is assumed. However, other utility functions can, of course, be dealt with. The reason for choosing risk neutrality is rather a question of mathematical tractability than economically motivated. These assumptions allow us to select a certain equivalent martingale measure $\tilde{\mathbb{P}} = \mathbb{P}_\tau \otimes \tilde{\mathbb{P}}_B$ under which risk neutral discounting with riskless constant interest rate r is feasible. Notice, that by considering regime switching risk over τ we implicitly fall into the class of stochastic volatility models. In the stochastic volatility literature risk neutrality towards volatility risk is a standard assumption as well.[8]

Furthermore, throughout the analysis it is assumed that the firm is the only supplier of the product line. Although this means that the firm acts as a monopolist, by specifying the demand parameter dynamics as a product life cycle we implicitly take account of competition since the firm faces the threat of being forced into the decay regime by exogenous competitive actions.

Once the firm is operating it has running costs depending on the installed capacity M and the current production rate Q_t of

$$c(t, \Theta_t, Q_t, M) = c_1 Q_t + \frac{c_2}{2M} Q_t^2 + c_3 M \qquad (6.6)$$

where c_1, c_2 and c_3 are positive constants, and the production rate Q_t is bounded to be in the interval $[0, M]$. The cost function needs some explanation. First, it is convex in Q_t which means that producing at full capacity is usually more expensive than somewhat below 100%. Second, there are two effects caused by different capacity levels. On the one hand, higher capacity lowers the production cost because the production rate is farther away from full capacity. On the other hand, higher capacity also means higher fixed cost of operation represented by the last summand of equation (6.6). Fixed cost may contain overhead cost, cost of capital or depreciation.

---

[8]See Hobson [95] for a good introduction and overview of stochastic volatility models.

In order to build the production facility the firm has to pay an initial investment
outlay of

$$H_I(t, \Theta_t, M) = c_4 M \tag{6.7}$$

where the positive constant c_4 denotes the cost of one unit of capacity. At the end of
the product life cycle T we assume that the production facility is sold for its scrap
value determined by the contraction cost function H_C below. After having built the
plant the firm has the option to either expand or contract its output capacity which
incurs another cost of

$$H_E(t, \Theta_t, M_0, M_1) = c_5(M_1 - M_0) + c_7 \tag{6.8}$$

in the case of capacity expansion $(M_1 > M_0)$ and

$$H_C(t, \Theta_t, M_0, M_1) = c_6(M_0 - M_1) + c_7 \tag{6.9}$$

in the case of capacity reduction $(M_1 < M_0)$. Again, c_5 and c_7 are assumed to
be positive and constant. c_5 denotes the unit cost of capacity expansion which is
usually higher than the initial unit cost c_4. Depending on the specific situation the
constant c_6 which stands for the unit disinvestment cost may take positive (unit cost
of disinvestment) or negative (scrap value of one unit of capacity) values. In any
case it is reasonable to assume that unit disinvestment cost are lower than initial
unit investment cost, i.e., $c_6 < c_4$. c_7 denotes the fixed cost of making a change in
capacity.

Finally, in the model considered here we assume that every capacity decision
takes effect immediately. Therefore we do not explicitly model the effect of time
lags on project values and optimal investment and operating policies.

In order to be able to compare the interaction effects of the different options
that are embedded into the project we are now going to develop the stochastic
control models and the graphical representations of the contingency structures for
several combinations of real options. First, we start with the project value for fixed
capacity but with variable output rates. Afterwards, the value of the timing and
intensity option is examined. The third case considers the flexible capacity option
under variable output rates while the last case examines all options together.

6.2.1 Static Project Value with Fixed Capacity

We start with the project value of the firm that has to invest immediately at time
0. The only source of flexibility that is open to the firm is to adjust its production
rate costlessly during the lifetime of the project and to choose the initial capacity.
Consequently, the firm is operating right from the start. There are only two states
to consider. The first is operating in the growth regime and the second is operating in the decay regime. Since switching from the growth regime to the decay
regime is exogenously governed, we have to introduce a new graphical modelling

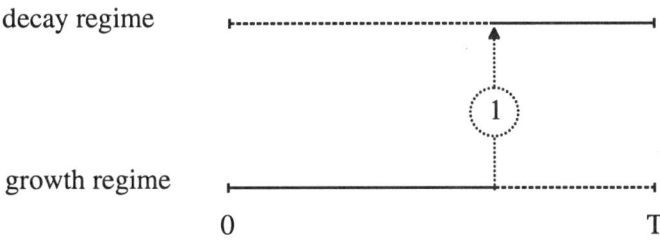

decay regime

growth regime

0   T

Figure 6.1: Graphical representation of the product life cycle model with modelling element exogenous random event

element called *exogenous random event*. Besides that, there are no other reversible or irreversible switches to consider. The corresponding graphical representation is displayed in Figure 6.1. The combination of dotted line and dotted circle represents an exogenous random event that cannot be influenced by the firm. Because the random event brings the system to exactly one different state, the number 1 is included into the circle. The rest of the graphical representation is as defined above.

The flexibility to adjust production rates Q_t is not included in the graph because it does not possess the nature of reversible or irreversible switches. To see this examine first the instantaneous cash flow function f. The instantaneous cash flow earned by the firm in both states is given by

$$
\begin{aligned}
f(t, \Theta_t, Q_t, M) &= Q_t P_t - c(t, \Theta_t, Q_t, M) & (6.10) \\
&= Q_t \frac{\Theta_t - Q_t}{\lambda} - c_1 Q_t - \frac{c_2}{2M} Q_t^2 - c_3 M \\
&= Q_t \left(\frac{\Theta_t}{\lambda} - c_1 \right) - Q_t^2 \left(\frac{1}{\lambda} + \frac{c_2}{2M} \right) - c_3 M
\end{aligned}
$$

where we have used the fact that under market clearance

$$
P_t = \frac{\Theta_t - Q_t}{\lambda} \qquad (6.11)
$$

holds. The ability of the firm to adjust production rates immediately at no extra cost is equivalent to maximizing the instantaneous cash flow function f with respect to $Q_t \in [0, M]$.[9] Performing the optimization simply means setting the first derivative

[9]In mathematical terms, the optimization of the project value with respect to the production rate is a continuous stochastic control problem. Maximizing the project value is equivalent to maximizing instantaneous cash flows because the resulting control strategy is Markovian, i.e., it does solely depend on the current state of the system and not on previous states. The latter kind is called a time (path) dependent continuous stochastic control problem and was for example considered by Cherian/Patel/Khripko [48] for the case of accumulating costs in an optimal mine extraction problem. We will not pursue this idea any further since in our application the idea for solving the continuous control problem is straightforward.

of f with respect to Q_t equal to zero:

$$\frac{\partial f(t, \Theta_t, Q_t, m)}{\partial Q_t} = \frac{\Theta_t}{\lambda} - c_1 - 2Q_t \left(\frac{1}{\lambda} + \frac{c_2}{2M} \right) \overset{!}{=} 0. \tag{6.12}$$

By noting that the second derivative of f with respect to Q_t is less than zero, this yields the optimal production rate Q_t^* as

$$Q_t^* = \frac{\Theta_t - \lambda c_1}{2 + \frac{\lambda c_2}{M}}. \tag{6.13}$$

Because production rates are limited to be in the interval $[0, M]$ and Θ_t is the only variable in equation (6.13), we have to check for which of its values Q_t^* becomes negative ($Q_t^* < 0$) or greater than capacity ($Q_t^* > M$), respectively. The first case yields

$$Q_t^* = \frac{\Theta_t - \lambda c_1}{2 + \frac{\lambda c_2}{M}} \quad < \quad 0 \tag{6.14}$$

$$\implies \quad \Theta_t \quad < \quad \lambda c_1$$

and the second

$$Q_t^* = \frac{\Theta_t - \lambda c_1}{2 + \frac{\lambda c_2}{M}} \quad > \quad M \tag{6.15}$$

$$\implies \quad \Theta_t \quad > \quad 2M + \lambda(c_1 + c_2).$$

Therefore, depending on the demand parameter value Θ_t the optimal production rate Q_t^* is given by

$$Q_t^* = \begin{cases} 0 & \text{for} & \Theta_t < \lambda c_1 & (I), \\ \frac{\Theta_t - \lambda c_1}{2 + \frac{\lambda c_2}{M}} & \text{for} & \lambda c_1 \le \Theta_t \le 2M + \lambda(c_1 + c_2) & (II), \\ M & \text{for} & \Theta_t > 2M + \lambda(c_1 + c_2) & (III). \end{cases} \tag{6.16}$$

The corresponding optimal cash flow function f for each of the three cases can be obtained by substituting Q_t^* back into equation (6.10). After some calculations we get

$$f(t, \Theta_t, M) = \begin{cases} -c_3 M & \text{for } (I), \\ \frac{1}{2 + \frac{\lambda c_2}{M}} \left(\frac{\Theta_t^2}{2\lambda} - c_1 \Theta_t + \frac{\lambda c_1^2}{2} \right) - c_3 M & \text{for } (II), \\ M \left(\frac{\Theta_t}{\lambda} - c_1 - \frac{c_2}{2} - c_3 \right) - \frac{M^2}{\lambda} & \text{for } (III). \end{cases} \tag{6.17}$$

Note, that for values of Θ_t less than λc_1 the optimal production rate is zero and the instantaneous cash flow negative. This actually means that the firm is in a state of costly suspension. Furthermore, it can be shown that the instantaneous cash flow function is continuous and increasing in Θ_t. However, the behavior of $f(, t, \Theta_t, M)$ on changes in M is ambiguous. For small values of Θ_t cash flows decrease with increasing capacity M reflecting the fact that more capacity is less profitable when demand is low due to the larger overhead cost of higher capacity. For large demand the situation is just turned around. High capacity yields higher instantaneous profits than low capacity.

We turn next to the project value of the investment opportunity without options. Let V_{Pg} be the project value in the growth regime and V_{Pd} the value of the project in the decay regime. We define the state of the system at time t by X_t. X_t is a vector consisting of the elements

$$X_t = \begin{pmatrix} t \\ \Theta_t \\ M \end{pmatrix} ; \tag{6.18}$$

t and Θ_t are as defined above. The action space $\mathcal{Z}$ of the firm now consists of the choice of the initial capacity M. It is reasonable to assume that there exist exogenous constraints for the possible range of capacity values. For example, there might be pollution constraints or constraints about the selected site and so on that limit the maximal installable capacity. On the other hand, there might as well be exogenous constraints on the minimal installable capacity. Think for example of subsidies that are only granted when a certain minimal size of the production facility is realized. Therefore, it is assumed that the initial installed capacity lies in the interval $\mathcal{Z} = [M_{min}, M_{max}]$. The initial value of X_t is given by $x = X_0 = (0, \Theta_0, M)$.

Let us start with the project value in the decay regime. The switching time τ is already revealed so that the state of the system is currently X_t with $t \geq \tau$. Then the only source of randomness is the Brownian motion B_t. The corresponding project value is consequently given by

$$V_{Pd}(X_t) = \tilde{\mathbb{E}}\left[\int_t^T e^{-r(s-t)} f(X_s) \, ds \,\Big|\, \mathcal{F}_t^d\right] \tag{6.19}$$

$$= \tilde{\mathbb{E}}_B\left[\int_t^T e^{-r(s-t)} f(X_s) \, ds \,\Big|\, \mathcal{F}_t^d\right].$$

This expression can again be expanded using Ito's lemma resulting in the following PDE for the project value in the decay regime,[10] i.e., for $\tau \leq t \leq T$,

$$\frac{1}{2}\sigma^2\Theta^2\frac{\partial^2 V_{Pd}}{\partial\Theta^2} + \mu_d\Theta\frac{\partial V_{Pd}}{\partial\Theta} + \frac{\partial V_{Pd}}{\partial t} - rV_{Pd} + f = 0. \tag{6.20}$$

For the project value in the growth regime the situation is slightly different. Because we do not know the switching time τ in advance we have to average over all possible

---

[10] The dependence on X_t was omitted for brevity.

values of $\tau > t$. Once switching from growth to decay has taken place the reward earned by the firm is just the value in the decay regime V_{Pd}. Thus, the value of the project at time t in the growth regime can be obtained by

$$V_{Pg}(X_t) \tag{6.21}$$
$$= \mathbb{E}_\tau \left[\tilde{\mathbb{E}}_B \left[\int_t^\tau e^{-r(s-t)} f(X_s)\, ds + e^{-r(\tau-t)} V_{Pd}(X_\tau) \,\middle|\, \mathcal{F}_t^g \right] \right].$$

In order to calculate the corresponding PDE for V_{Pg} we have to remember that the probability of switching to the decay regime in the next small time interval dt is given by $l(t)\,dt$. Hence, with probability $l(t)\,dt$ the value of the project will drop to V_{Pd}. With a probability of $1 - l(t)\,dt$ the process stays in the growth regime and the value function can be expanded using Ito's lemma. Furthermore, in the time interval $(t, t+dt)$ the firm earns an instantaneous cash flow of $f(X_t)dt$. Thus, we get[11]

$$V_{Pg}(X_t) \tag{6.22}$$
$$= f(X_t)\, dt + l(t)\, dt\, V_{Pd}(X_t) + ((1 - l(t)\, dt)\, e^{-r dt}$$
$$\times \left[V_{Pg}(X_t) + \left[\frac{1}{2}\sigma^2\Theta_t^2 \frac{\partial^2 V_{Pg}(X_t)}{\partial\Theta^2} + \mu_d\Theta_t \frac{\partial V_{Pg}(X_t)}{\partial\Theta} + \frac{\partial V_{Pd}(X_t)}{\partial t} \right] dt \right].$$

By noting that $e^{-r dt} \approx 1 - r dt$ for dt small and simplifying the expression, the project value in the growth regime satisfies the PDE[12]

$$\frac{1}{2}\sigma^2\Theta^2 \frac{\partial^2 V_{Pg}}{\partial\Theta^2} + \mu_d\Theta \frac{\partial V_{Pg}}{\partial\Theta} + \frac{\partial V_{Pd}}{\partial t} \tag{6.23}$$
$$- (r + l(t))V_{Pg} + f + l(t)\, V_{Pd} = 0$$

for all $0 \le t < \tau$.

At terminal time T both in the growth regime and in the decay regime the initially installed capacity M is sold for its scrap value. Therefore we have to add the two terminal conditions

$$V_{Pg}(T, \Theta_T, M) = -H_C(T, \Theta_T, M, 0) = -c_6 M - c_7, \tag{6.24}$$
$$V_{Pd}(T, \Theta_T, M) = -H_C(T, \Theta_T, M, 0) = -c_6 M - c_7, \tag{6.25}$$

admitting positive as well as negative salvage values.

Unfortunately, neither the PDE (6.20) for the project value in the decay regime V_{Pd} nor the PDE (6.23) for the project value in the growth regime V_{Pg} can be solved analytically. Neither admits a closed form solution because the cash flow function is a piecewise defined function of Θ. We will show how to solve the PDEs numerically in Section 6.3.

---

[11] See the derivation of the generalized Ito formula in Section 3.2.2.
[12] The dependence on X_t was omitted for brevity.

In order to enable management to judge the profitability of the (static) investment project, we need to determine the value of the project in the growth regime at time zero. Since management can choose the initial capacity M_0 it has to find M_0 such that the expected profit of the firm minus the initial investment outlay necessary to build a plant of capacity M_0 is maximized, i.e.,

$$\max_{M \in [M_{min}, M_{max}]} \left\{ V_{Pg}(0, \Theta_0, M) - c_4 M \right\} \quad > \quad 0. \tag{6.26}$$

If this objective function is negative, the project is rejected. This corresponds to the classical net present value rule of rejecting investments with negative net present value. The reason why the NPV rule is applicable in this case is that management has no future flexibility that can be interpreted as real option. Its production strategy is a priori fixed by selecting the optimal initial capacity level and the already determined optimal production rate which is independent of time.

6.2.2 Timing and Intensity

If the firm possesses in addition the strategic option to choose the timing and intensity of investment, the optimal strategy cannot be judged using an NPV rule. A real option analysis has to be performed in order to determine the expanded project value of the investment opportunity.

Starting again with the graphical contingency structure representation there are three states that can be identified. In addition, to the states operating in growth regime and operating in decay regime the firm can as well be in a state of waiting. When the firm is waiting it earns no cash flow. Once the demand parameter reaches a certain trigger level Θ_ϑ^* at time ϑ the firm will start to operate at an optimal capacity level M_ϑ^*. In order to build the production facility, switching costs of $c_4 M_\vartheta^*$ are incurred. The corresponding graphical representation of the decision situation displayed in Figure 6.2 also takes account of the irreversibility of the timing and intensity option.

Before we formulate the corresponding optimal stopping problem we have to distinguish the order of the regime switch and the decision to exercise the timing and intensity option. In Figure 6.2 it is implicitly assumed that the decision to invest is made while the demand parameter is in the growth regime, i.e., $\vartheta < \tau$. This will be most likely the case because there is much less incentive to invest when the demand parameter is already in the decay regime. However, we have to take this case into account as well. Consequently, as for the project value we have to consider two option values of waiting. The first one is the option value of waiting in the growth regime denoted by $V_{Wg}(t, \Theta_t, 0)$ while the second one represents the option value in the decay regime denoted by $V_{Wd}(t, \Theta_t, 0)$.

The value of the timing and intensity option at time t in the decay regime is

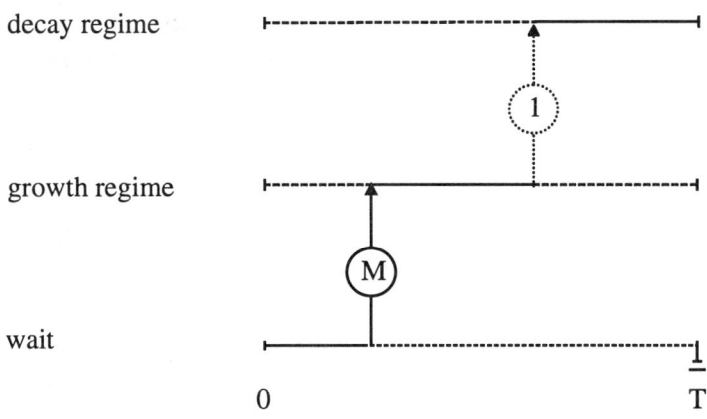

Figure 6.2: Product life cycle model with timing and intensity option

therefore given by the solution of the generalized timing option[13]

$$V_{Wd}(X_t) \tag{6.27}$$

$$= \sup_{(\vartheta, M) \in [0,T] \times \mathcal{Z}} \tilde{\mathbb{E}}_B \left[e^{-r(\vartheta - t)} V_{Pd}(X_\vartheta) - e^{-r(\vartheta - t)} c_4 M \right].$$

Note that in this case $\tau \leq t \leq \vartheta$ holds.

In order to derive the quasi-variational inequalities of the above optimization problem we first expand V_{Wd} using Itô's lemma, which results in the PDE

$$\frac{1}{2}\sigma^2\Theta^2 \frac{\partial^2 V_{Wd}}{\partial \Theta^2} + \mu_d \Theta \frac{\partial V_{Wd}}{\partial \Theta} + \frac{\partial V_{Wd}}{\partial t} - r V_{Wd} = 0 \tag{6.28}$$

where the state dependence of the functions was again omitted. Next we define the maximum operator

$$\mathcal{M}^\vartheta V_{Wd}(t, \Theta_t, 0) := \max_{M \in \mathcal{Z}} \{ V_{Pd}(t, \Theta_t, M) - c_4 M \} \tag{6.29}$$

which means that, at each time t with current demand parameter realization Θ_t, management has to check whether it is better off by starting operations with a capacity level M or not. If the option value of further waiting is greater than the project value net cost of installing capacity, the firm will continue in its idle state. If on the other hand the demand parameter level Θ_t rises high enough at time t and reaches the trigger level $\Theta_d^\vartheta(t)$ the firm chooses the corresponding optimal capacity

---

[13]From now on we omit the dependence of the conditional expectation on the filtration $\mathcal{F}_t$ at time t for brevity.

$M_d^\vartheta(t)$ that maximizes its project value in operation minus investment cost. This consideration gives rise to the continuation region $\mathcal{C}_d^\vartheta$ in the decay regime

$$\mathcal{C}_d^\vartheta := \left\{ (t, \Theta_t) \in [0, T] \times \mathbb{R}_+;\ V_{Wd} > \mathcal{M}^\vartheta V_{Wd} \right\} \qquad (6.30)$$

$$= \left\{ (t, \Theta_t) \in [0, T] \times \mathbb{R}_+;\ \Theta_t < \Theta_d^\vartheta(t) \right\}.$$

Consequently, the quasi-variational inequalities of the stopping problem for all $X_t \in [0, T] \times \mathbb{R}_+ \times \{0\}$ are:

$$\frac{1}{2} \sigma^2 \Theta^2 \frac{\partial^2 V_{Wd}}{\partial \Theta^2} + \mu_d \Theta \frac{\partial V_{Wd}}{\partial \Theta} + \frac{\partial V_{Wd}}{\partial t} - r V_{Wd} \leq 0 \qquad (6.31)$$

$$V_{Wd} \geq \mathcal{M}^\vartheta V_{Wd}$$

One of the inequalities is an equality.

In order to find a solution to this set of quasi-variational inequalities, first note that in the continuation region $\mathcal{C}_d^\vartheta$ the first inequality must be an equality. Moreover, at the boundary of the continuation region $\partial \mathcal{C}_d^\vartheta$ the second inequality must be an equality. Together with the condition that V_{Wd} is stochastically C^2, this results in the value matching condition[14] for the free boundary $\Theta_d^\vartheta(t)$ and the corresponding optimal capacity $M_d^\vartheta(t)$ in the decay regime for all $t \in [0, T]$,

$$V_{Wd}(t, \Theta_d^\vartheta(t), 0) = V_{Pd}(t, \Theta_d^\vartheta(t), M_d^\vartheta(t)) - c_4 M_d^\vartheta(t). \qquad (6.32)$$

Finally, we have to add the terminal condition that at time T the option to invest is worthless. Hence,

$$V_{Wd}(T, \Theta_T, 0) = 0. \qquad (6.33)$$

As already for the project value there exists no closed form solution for the option value in the decay regime. However, we will be able to solve for the option value numerically such that the resulting value function is a solution to the system of quasi-variational inequalities (6.31). This makes the verification theorem of Section 3.4.2 applicable and proves the optimality of the results.

The value of the timing and intensity option at time t in the growth regime can be found similarly as in the decay regime. The corresponding optimal stopping problem is now slightly different because we have to take the chance of regime switching into consideration. This is reflected by the expectation $\tilde{\mathbb{E}}$ instead of $\tilde{\mathbb{E}}_B$ as above:

$$V_{Wg}(X_t) = \sup_{(\vartheta, M) \in [0, T] \times \mathcal{Z}} \tilde{\mathbb{E}} \left[e^{-r(\vartheta - t)} V_{Pg}(X_\vartheta) - e^{-r(\vartheta - t)} c_4 M \right]. \qquad (6.34)$$

---

[14]As it will turn out in the numerical analysis below we need only the value matching condition to numerically determine the boundary of the continuation region $\partial \mathcal{C}_d^\vartheta$. Therefore the smooth pasting condition is omitted.

If we want to expand this expression we cannot be sure whether the option to invest is exercised before (at $\vartheta < \tau$) or after (at $\vartheta \geq \tau$) regime switching has taken place. For this reason we proceed again as for the project value in the growth regime in order to derive the following PDE for the value of the timing and intensity option in the growth regime

$$\frac{1}{2}\sigma^2\Theta^2\frac{\partial^2 V_{Wg}}{\partial\Theta^2} + \mu_g\Theta\frac{\partial V_{Wg}}{\partial\Theta} + \frac{\partial V_{Wg}}{\partial t} \\ -(r+l(t))V_{Wg} + l(t)V_{Wd} = 0. \tag{6.35}$$

The maximum operator is defined as above

$$\mathcal{M}^\vartheta V_{Wg}(t,\Theta_t,0) := \max_{M\in\mathcal{Z}}\left\{V_{Pg}(t,\Theta_t,M) - c_4\,M\right\} \tag{6.36}$$

as well as the continuation region in the growth regime

$$\begin{aligned} \mathcal{C}_g^\vartheta &:= \left\{(t,\Theta_t)\in[0,\tau)\times\mathbb{R}_+;\ V_{Wg} > \mathcal{M}^\vartheta V_{Wg}\right\} \\ &= \left\{(t,\Theta_t)\in[0,\tau)\times\mathbb{R}_+;\ \Theta_t < \Theta_g^\vartheta(t)\right\} \end{aligned} \tag{6.37}$$

with the optimal capacity $M_g^\vartheta(t)$. Consequently, the quasi-variational inequalities in the growth regime for all $X_t \in [0,\tau)\times\mathbb{R}_+\times\{0\}$ are

$$\begin{aligned} \frac{1}{2}\sigma^2\Theta^2\frac{\partial^2 V_{Wg}}{\partial\Theta^2} + \mu_g\Theta\frac{\partial V_{Wg}}{\partial\Theta} + \frac{\partial V_{Wg}}{\partial t} \\ -(r+l(t))V_{Wg} + l(t)V_{Wd} &\leq 0 \\ V_{Wg} &\geq \mathcal{M}^\vartheta V_{Wg} \end{aligned} \tag{6.38}$$

One of the inequalities is an equality.

Furthermore, the value matching condition for the optimal trigger demand level $\Theta_g^\vartheta(t)$ and the optimal capacity to install $M_g^\vartheta(t)$ at time t in the growth regime is given by

$$V_{Wg}(t,\Theta_g^\vartheta(t),0) = V_{Pg}(t,\Theta_g^\vartheta(t),M_g^\vartheta(t)) - c_4\,M_g^\vartheta(t). \tag{6.39}$$

Finally, the terminal condition is

$$V_{Wg}(T,\Theta_T,0) = 0. \tag{6.40}$$

Note, that because the optimization problem is time inhomogeneous the optimal entry triggers $\Theta_g^\vartheta(t)$ and $\Theta_d^\vartheta(t)$ as well as the optimal entry capacity levels $M_g^\vartheta(t)$ and $M_d^\vartheta(t)$ in the growth and decay regime change over time. This is one of the main differences from the time homogeneous problems discussed in the above chapters.

6.2.3 Flexible Capacity

The next case to consider is the option to adjust capacity while the firm is operating up front. In this case the firm has the option to choose the initial capacity level M_0 that maximizes its value. In addition, the firm has the option to adjust capacity in a range between $\mathcal{Z} = [M_{min}, M_{max}]$ upwards or downwards by paying the switching costs H_E or H_C whenever upward or downward changes of the demand parameter during the life time of the production facility are large enough. Intuitively, the flexibility to adjust capacity is very desirable in the presence of large demand fluctuations because it enables management to cut the fixed cost incurred by maintaining large capacity at low demand. On the other hand if demand increases beyond expectation, the ability to expand the production facility can be the source of additional profits.

Since capacity adjustments are costly reversible the described situation can be modelled as a generalized switching option. Note that in case $M_{min} = 0$ the option to abandon is implicitly included in the model. Depending on the current state $X_t = (t, \Theta_t, M_t)$ the firm earns an instantaneous cash flow given by equation

$$f(t, \Theta_t, M_t) = \begin{cases} -c_3\,M_t & \text{for } (I), \\ \frac{1}{2+\frac{\lambda c_2}{M_t}}\left(\frac{\Theta_t^2}{2\lambda} - c_1\Theta_t + \frac{\lambda c_1^2}{2}\right) - c_3 M_t & \text{for } (II), \\ M_t\left(\frac{\Theta_t}{\lambda} - c_1 - \frac{c_2}{2} - c_3\right) - \frac{M_t^2}{\lambda} & \text{for } (III), \end{cases} \qquad (6.41)$$

which is the same as (6.17) with M_t now time dependent. The initial investment cost function H_I and the switching cost functions H_E and H_C are given by equations (6.7) to (6.9), respectively. The resulting graphical representation of the investment problem is displayed in Figure 6.3.

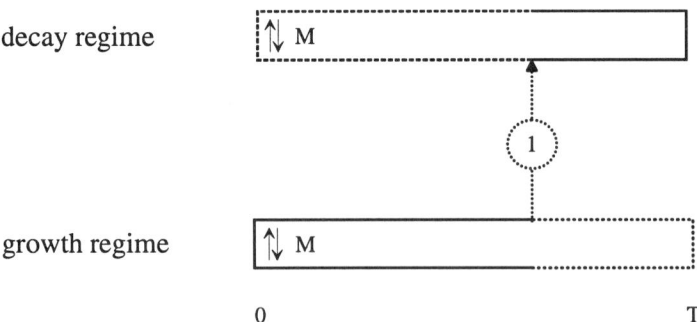

decay regime

growth regime

0   T

Figure 6.3: Product life cycle model with flexible capacity

As can be seen from the figure we have again to distinguish the value of the firm with flexible capacity in the growth regime V_{Fg} and in the decay regime V_{Fd}. Before formulating the optimization problem for both cases we have to define

Before formulating the optimization problem for both cases we have to define feasible impulse control strategies. Again an impulse control strategy consists of a finite sequence of stopping times $\theta_i^d \in [t, T]$ and corresponding impulse controls, i.e., upward or downward adjustments of capacity, $\zeta_i^d = M_{\theta_i^d} = M_i^d \in \mathcal{Z}$ when the firm operates at time $\theta_0 = t$ with capacity $\zeta_0 = M_t$. Let $w_d \in \mathcal{W}_d$ be an admissible impulse control strategy in the decay regime. Then the optimization problem of the firm already in the decay regime with $X_t = (t, \Theta_t, M_t)$ that has to be solved is

$$V_{Fd}(X_t) = \sup_{w^d \in \mathcal{W}^d} \tilde{\mathbb{E}}_B \left[\int_t^T e^{-r(s-t)} f(X_s)\, ds \right. \tag{6.42}$$

$$\left. - \sum_{i:t \le \theta_i^d \le T} e^{-r(\theta_i^d - t)} H_{E/C}(\theta_i^d, \Theta_{\theta_i^d}, M_{i-1}^d, M_i^d) \right]$$

where $H_{E/C}$ denotes the switching cost of capacity expansion H_E if $M_{i-1}^d < M_i^d$ and H_C if $M_{i-1}^d > M_i^d$. In order to determine the set of quasi-variational inequalities for the optimization problem (6.42), first define the maximum operator

$$\mathcal{M}V_{Fd}(t, \Theta_t, M) \tag{6.43}$$

$$:= \max_{M^* \in \mathcal{Z}\backslash\{M\}} \left\{ V_{Fd}(t, \Theta_t, M^*) - H_{E/C}(t, \Theta_t, M, M^*) \right\}$$

$$= \max \left\{ \max_{M^* \in \mathcal{Z}, M^* > M} \left\{ V_{Fd}(t, \Theta_t, M^*) - H_E(t, \Theta_t, M, M^*) \right\}, \right.$$

$$\left. \max_{M^{**} \in \mathcal{Z}, M^{**} < M} \left\{ V_{Fd}(t, \Theta_t, M^{**}) - H_C(t, \Theta_t, M, M^{**}) \right\} \right\}.$$

This means that whenever changing the current level of capacity is optimal the new capacity level is chosen such that it maximizes the new project value net switching cost over all possible upward or downward adjustments. Next, we expand the value function as above. Together, this yields the following set of quasi-variational inequalities in the decay regime for $X_t \in [0, T] \times \mathbb{R}_+ \times [M_{min}, M_{max}]$:

$$\frac{1}{2}\sigma^2\Theta^2 \frac{\partial^2 V_{Fd}}{\partial \Theta^2} + \mu_d \Theta \frac{\partial V_{Fd}}{\partial \Theta} + \frac{\partial V_{Fd}}{\partial t} - r V_{Fd} + f \le 0$$

$$V_{Fd} \ge \mathcal{M}V_{Fd} \tag{6.44}$$

One of the inequalities is an equality.

Since M is assumed to be a continuous variable, the corresponding impulse control problem is different from the entry and exit models or a simple stopping model like the option to wait. For the option to wait to invest we had exactly one system of quasi-variational inequalities, and for the entry and exit decision we had two systems of quasi-variational inequalities—one for each state. However, M can take values in the interval $[M_{min}, M_{max}]$ continuously. Therefore, for all of the infinitely many Ms we have to solve the infinite set of quasi-variational inequalities simultaneously. In order to solve such an infinite system numerically there is no way out but to discretize the space of possible capacity levels which reduces the problem to a finite one.

The next important question to answer is what particular form the continuation region for the impulse control problem takes. By definition the continuation region is given by

$$\mathcal{C}_d := \left\{ (t, \Theta_t, M_t) \in [0, T] \times \mathbb{R}_+ \times \mathcal{Z}; \ V_{Fd} > \mathcal{M} V_{Fd} \right\}. \tag{6.45}$$

While this does not help much, observe that the only strategic variable that can be adjusted to changes in demand is the capacity level M. The cash flow function $f(X_t)$ is increasing in Θ_t. As a consequence, the project value is increasing in Θ_t with M constant. Moreover, since for low demand the cash flow function is decreasing in M, we conjecture that for a sufficiently low demand level Θ_d^{**} a downward adjustment of capacity to the level $M_d^{**} < M$ is optimal. On the other hand, if demand turns out to be high enough and reaches a trigger level Θ_d^*, the higher reward earned by operating at a higher capacity may justify incurring the switching cost of expanding production to $M_d^* > M$. Although we are not able to analytically prove this result the numerical analysis in Section 6.3, which is not restricted to this special intervention strategy, suggests that this particular strategy is chosen as optimal. Therefore, the continuation region for the impulse control problem has the particular form

$$\mathcal{C}_d := \tag{6.46}$$
$$\left\{ (t, \Theta_t, M_t) \in [0, T] \times \mathbb{R}_+ \times \mathcal{Z}; \ \Theta_d^{**}(t, M_t) < \Theta_t < \Theta_d^*(t, M_t) \right\}$$

with the corresponding lower and upper target capacity levels $M_d^{**}(t, M_t)$ and $M_d^*(t, M_t)$, respectively. Hence, in the continuation region the first inequality of (6.44) is an equality and the system evolves freely without any capacity adjustments. Once the demand parameter reaches the upper or lower boundary of the continuation region $\partial \mathcal{C}_d$, the second inequality is an equality and, according to the definition of the maximum operator $\mathcal{M}$ in (6.43), this yields the value matching conditions for the two free-boundaries $\Theta_d^{**}(t, M_t)$ and $\Theta_d^*(t, M_t)$ for all $t \in [0, T]$ and $M_t \in \mathcal{Z}$:

$$V_{Fd}(t, \Theta_d^{**}, M_t) \ = \ V_{Fd}(t, \Theta_d^{**}, M_d^{**}) - H_C(t, \Theta_d^{**}, M_t, M_d^{**}), \tag{6.47}$$
$$V_{Fd}(t, \Theta_d^*, M_t) \ = \ V_{Fd}(t, \Theta_d^*, M_d^*) - H_E(t, \Theta_d^*, M_t, M_d^*). \tag{6.48}$$

Finally, as for the project value above, we add the terminal condition

$$V_{Fd}(T, \Theta_T, M_T) \ = \ -H_C(T, \Theta_T, M_T, 0) \tag{6.49}$$
$$= \ -c_6 M_T - c_7, \ \forall \, \Theta_T \in \mathbb{R}_+, \ M_T \in \mathcal{Z}.$$

Note that the expansion and contraction trigger demand levels $\Theta_d^*(t, M_t)$ and $\Theta_d^{**}(t, M_t)$, as well as the optimal expansion and contraction capacities $M_d^*(t, M_t)$

and $M_d^{**}(t, M_t)$, now depend not only on time but also on the currently installed capacity M_t. This means that for each time t and each capacity level M_t there exists another optimal expansion and contraction policy. In addition, there might be capacity levels M_t for which expansion or contraction is never optimal. For example, if capacity has already reached its lowest level, i.e., $M_t = M_{min}$, then there exists no further opportunity to contract and the contraction trigger is not defined. The same holds true for the upper capacity boundary M_{max}. There might also be cases where the contraction or expansion trigger is zero or infinity. In both cases the boundaries of the continuation region that trigger contraction or expansion cannot be attained. All of these mentioned cases have to be considered in the numerical solution procedure. Furthermore, if $M_{min} = M_{max} = M$ there exists no flexibility to adjust capacity and we recover the (static) project $V_{Pd}(t, \Theta_t, M)$.

In the growth regime an impulse control is defined in a slightly more complicated way. For the growth regime we have that $t < \tau$. Consequently, if the current state of the production system is $X_t = (t, \Theta_t, M_t)$, then an impulse control for the remaining time to the end of the product life cycle T is defined as the finite sequence of stopping times $t \leq \theta_i^g < \tau$ for $i = 1, \dots, k$ and corresponding impulse controls $\zeta_i^g = M_i^g$ and, additionally, the impulse control strategy w_d in the decay regime with $\theta_0^d = \tau$, Θ_τ and $M_0^d = M_\tau$ as starting values. Hence, an impulse control strategy w is defined as

$$
\begin{aligned}
w \quad &:= \quad (\theta_1^g, \dots, \theta_k^g, \tau, \theta_1^d, \dots, \theta_k^d; M_1^g, \dots, M_k^g, M_0^d, M_1^d, \dots, M_k^d) \\
&= \quad (w_g | (\tau, M_0^d) | w_d) \in \mathcal{W} \qquad\qquad\qquad (6.50)
\end{aligned}
$$

with θ_1^g the first impulse control time after t, θ_k^g the latest impulse control time before τ, θ_k^d the latest impulse control time before T and $w_g \in \mathcal{W}_g$ the impulse control in the growth regime.. Then the corresponding impulse control problem for the value of production flexibility in the growth regime is given by

$$
\begin{aligned}
V_{Fg}(X_t) \quad = \quad &\sup_{w \in \mathcal{W}} \tilde{\mathbb{E}}\Bigg[\int_t^T e^{-r(s-t)} f(X_s)\, ds \qquad\qquad\qquad\qquad (6.51) \\
&- \sum_{i:t \leq \theta_i^g \leq \tau} e^{-r(\theta_i^g - t)} H_{E/C}(\theta_i^g, \Theta_{\theta_i^g}, M_{i-1}^g, M_i^g) \\
&- \sum_{i:\tau \leq \theta_i^d \leq T} e^{-r(\theta_i^d - t)} H_{E/C}(\theta_i^d, \Theta_{\theta_i^d}, M_{i-1}^d, M_i^d) \Bigg] \\
= \quad &\mathbb{E}_\tau \Bigg[\sup_{w_g \in \mathcal{W}_g} \tilde{\mathbb{E}}_B \Bigg[\int_t^\tau e^{-r(s-t)} f(X_s)\, ds \\
&- \sum_{i:t \leq \theta_i^g \leq \tau} e^{-r(\theta_i^g - t)} H_{E/C}(\theta_i^g, \Theta_{\theta_i^g}, M_{i-1}^g, M_i^g) \\
&+ e^{-r(\tau - t)} V_{Fd}(X_\tau) \Bigg] \Bigg].
\end{aligned}
$$

Using the same maximum operator as in (6.42) and expanding (6.51) using Ito's lemma yields the system of quasi-variational inequalities which the value function in the growth regime $V_{Fd}(X_t)$ has to satisfy for all $X_t = (t, \Theta_t, M_t) \in [0, \tau) \times \mathbb{R}_+ \times \mathcal{Z}$:

$$
\begin{aligned}
\frac{1}{2}\sigma^2 \Theta^2 \frac{\partial^2 V_{Fg}}{\partial \Theta^2} + \mu_g \Theta \frac{\partial V_{Fg}}{\partial \Theta} + \frac{\partial V_{Fg}}{\partial t} & \\
- (r + l(t))V_{Fg} + l(t)V_{Fd} + f & \leq 0 \\
V_{Fg} & \geq \mathcal{M}\, V_{Fg}
\end{aligned}
\tag{6.52}
$$

One of the inequalities is an equality.

The same logic for the derivation of the continuation region applies to the growth regime as well. Therefore, we have

$$
\mathcal{C}_g := \left\{ (t, \Theta_t, M_t) \in [0, \tau) \times \mathbb{R}_+ \times \mathcal{Z};\; \Theta_g^{**}(t, M_t) < \Theta_t < \Theta_g^{*}(t, M_t) \right\}
\tag{6.53}
$$

with optimal new expansion and contraction capacities $M_g^{*}(t, M_t)$ and $M_g^{**}(t, M_t)$. However, there is one main difference from the continuation region in the decay regime. Since, the growth regime is left at the random switching time τ, the continuation region $\mathcal{C}_g$ is left as well. At the time instant τ the decay regime takes over there might be combinations of $X_\tau = (\tau, \Theta_\tau, M_k^g)$ which would have been in the continuation region of the growth regime $\mathcal{C}_g$ but which are *not* in the continuation region of the decay regime $\mathcal{C}_d$. This means that immediately after the switching time τ is revealed and the production system is in the decay regime, switching to another capacity level M_0^d—most probably to a lower one—is triggered. For this reason we distinguished M_k^g and M_0^d which otherwise are equal.

The two value matching conditions resulting from the maximum operator at the boundaries of the continuation region $\mathcal{C}_g$ follow with

$$
\begin{aligned}
V_{Fg}(t, \Theta_g^{**}, M_t) &= V_{Fg}(t, \Theta_g^{**}, M_g^{**}) - H_C(t, \Theta_g^{**}, M_t, M_g^{**}), &(6.54) \\
V_{Fg}(t, \Theta_g^{*}, M_t) &= V_{Fg}(t, \Theta_g^{*}, M_g^{*}) - H_E(t, \Theta_g^{*}, M_t, M_g^{*}), &(6.55)
\end{aligned}
$$

in addition to the terminal condition[15]

$$
\begin{aligned}
V_{Fg}(T, \Theta_T, M_T) &= -H_C(T, \Theta_T, M_T, 0) &(6.56) \\
&= -c_6 M_T - c_7
\end{aligned}
$$

for all $\Theta_T \in \mathbb{R}_+$ and $M_T \in \mathcal{Z}$. This concludes the mathematical analysis of production flexibility.

---

[15]The terminal condition could have been omitted in the mathematical presentation because, by assuming the truncated normal distribution for the regime switching time τ, the probability of the product life cycle still being in the growth regime at the terminal time T is zero. However, we need this boundary condition for the numerical analysis later in the chapter.

6.2.4   Timing, Intensity and Flexible Capacity

Since the firm is not committed to invest immediately we have to consider the timing and intensity option in combination with the flexible capacity option. All important steps to describe the model have already been taken. The graphical decomposition representation is just a combination of the timing and intensity option and the generalized switching options in each of the two regimes as can be seen in Figure 6.4.

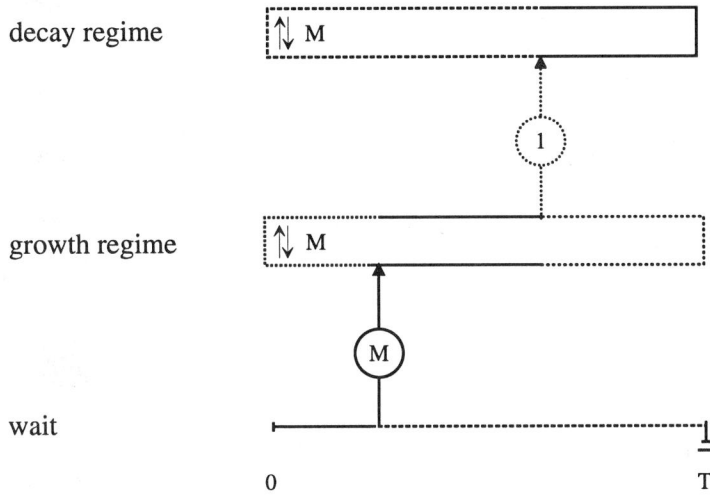

Figure 6.4: Product life cycle model with combined timing, intensity and flexible capacity

From a mathematical point of view the only difference from the timing and intensity option above is that now the exercise decision is whether to exchange the value of the option to invest with a) the value of the project with flexible capacity minus investment cost instead of b) the project value with fixed capacity minus investment cost. For example, in the decay regime the corresponding optimization problem reads

$$V_{WFd}(X_t) = \sup_{(\vartheta, M) \in [0,T] \times \mathcal{Z}} \tilde{\mathbb{E}}_B \left[e^{-r(\vartheta - t)} V_{Fd}(X_\vartheta) - e^{-r(\vartheta - t)} c_4 M \right]. \quad (6.57)$$

This is the same as (6.27) with the static project value V_{Pd} replaced by the project value with flexible capacity V_{Fd}. In the growth regime we get the optimization problem

$$V_{WFg}(X_t) = \sup_{(\vartheta, M) \in [0,T] \times \mathcal{Z}} \tilde{\mathbb{E}} \left[e^{-r(\vartheta - t)} V_{Fg}(X_\vartheta) - e^{-r(\vartheta - t)} c_4 M \right]. \quad (6.58)$$

The maximum operators are, similar to (6.29),

$$\mathcal{M}_F^{\vartheta} V_{WFd}(t, \Theta_t, 0) := \max_{M \in \mathcal{Z}} \{V_{Fd}(t, \Theta_t, M) - c_4 M\}, \qquad (6.59)$$

$$\mathcal{M}_F^{\vartheta} V_{WFg}(t, \Theta_t, 0) := \max_{M \in \mathcal{Z}} \{V_{Fg}(t, \Theta_t, M) - c_4 M\}. \qquad (6.60)$$

This results in the continuation regions for the decay and growth regime

$$\mathcal{C}_{Fd}^{\vartheta} := \left\{(t, \Theta_t) \in [0, T] \times \mathbb{R}_+; \; \Theta_t < \Theta_{Fd}^{\vartheta}(t)\right\}, \qquad (6.61)$$

$$\mathcal{C}_{Fg}^{\vartheta} := \left\{(t, \Theta_t) \in [0, \tau] \times \mathbb{R}_+; \; \Theta_t < \Theta_{Fg}^{\vartheta}(t)\right\}, \qquad (6.62)$$

where $\Theta_{Fd}^{\vartheta}(t)$ and $\Theta_{Fg}^{\vartheta}(t)$ denote the trigger prices of the timing and intensity options in the presence of flexible capacity in the decay and growth regime, respectively. In addition, we have the corresponding optimal capacities $M_{Fd}^{\vartheta}(t)$ and $M_{Fg}^{\vartheta}(t)$.

The system of quasi-variational inequalities in the decay regime is for all $X_t \in [0, T] \times \mathbb{R}_+ \times \mathcal{Z}$,

$$\frac{1}{2}\sigma^2 \Theta^2 \frac{\partial^2 V_{WFd}}{\partial \Theta^2} + \mu_d \Theta \frac{\partial V_{WFd}}{\partial \Theta} + \frac{\partial V_{WFd}}{\partial t}$$
$$-r V_{WFd} \leq 0$$
$$V_{WFd} \geq \mathcal{M}_F^{\vartheta} V_{WFd}, \qquad (6.63)$$

one of the inequalities is an equality,

and in the growth regime for all $X_t \in [0, \tau] \times \mathbb{R}_+ \times \mathcal{Z}$,

$$\frac{1}{2}\sigma^2 \Theta^2 \frac{\partial^2 V_{WFg}}{\partial \Theta^2} + \mu_g \Theta \frac{\partial V_{WFg}}{\partial \Theta} + \frac{\partial V_{WFg}}{\partial t}$$
$$- (r + l(t)) V_{WFg} + l(t) V_{WFd} \leq 0$$
$$V_{WFg} \geq \mathcal{M}_F^{\vartheta} V_{WFg} \qquad (6.64)$$

One of the inequalities is an equality.

The value matching conditions for both regimes are given by

$$V_{WFd}(t, \Theta_{Fd}^{\vartheta}(t), 0) = V_{Fd}(t, \Theta_{Fd}^{\vartheta}(t), M_{Fd}^{\vartheta}(t)) - c_4 M_{Fd}^{\vartheta}(t), \quad (6.65)$$

$$V_{WFg}(t, \Theta_{Fg}^{\vartheta}(t), 0) = V_{Fg}(t, \Theta_{Fg}^{\vartheta}(t), M_{Fg}^{\vartheta}(t)) - c_4 M_{Fg}^{\vartheta}(t), \quad (6.66)$$

where $t \in [0, T]$. Finally, we have the terminal conditions

$$V_{WFd}(T, \Theta_T, M_T) = 0, \qquad (6.67)$$

$$V_{WFg}(T, \Theta_T, M_T) = 0, \qquad (6.68)$$

for all $\Theta_T \in \mathbb{R}_+$ and $M_T \in \mathcal{Z}$.

In order to compare the values and the optimal operating policies of all of the developed models we briefly summarize the nomenclature of the involved variables in the growth regime in the following table:[16]

---

[16]The corresponding variable names in the decay regime are given by g replaced by d.

Model	Value Function	Trigger Price	Capacity Level
Static Project	$V_{Pg}(X_t)$	–	M_0
Timing and Intensity	$V_{Wg}(X_t)$	$\Theta_g^{\vartheta}(t)$	$M_g^{\vartheta}(t)$
Flexible Capacity	$V_{Fg}(X_t)$	$\Theta_g^{**}(t, M_t)$	$M_g^{**}(t, M_t)$
		$\Theta_g^{*}(t, M_t)$	$M_g^{*}(t, M_t)$
Timing, Intensity and Flexibility	$V_{WFg}(X_t)$	$\Theta_{Fg}^{\vartheta}(t)$	$M_{Fg}^{\vartheta}(t)$

Table 6.1: Summary of the variables' names in the growth regime

6.3 Numerical Solution Techniques

Before presenting the numerical results of the product life cycle model and its associated option values, we first address the question of numerical solution techniques that can be applied to value arbitrary combinations of generalized timing and switching options in real option interaction models. In the financial options literature a number of numerical methods have been used to value European and American style options. These methods can be broadly distinguished as:[17]

- methods that approximate the underlying stochastic variables (e.g., binomial trees and Monte Carlo simulation),

- methods that approximate the partial differential equations (e.g., finite differences), and

- analytical approximation techniques.

The first two approaches are used for numerical analyses of real options. So far, analytical approximation techniques have only been applied to simple option structures, as for example simple finite maturity American options,[18] Due to the complexity of the generalized timing and intensity options, finding analytical approximations is certainly the rare exception. Therefore these valuation techniques seem to be not appropriate for valuing real option interactions.

The most intuitive methods in pricing real and financial options are probably triangular lattice techniques of which binomial trees are a special case. The basic idea of binomial trees is to partition the time interval of interest in small time steps

---

[17]An excellent treatment of numerical methods applied to financial options can be found in the book of Hull [102], Chapter 15, and the references therein.

[18]See e.g., Carr/Faguet [45].

and determine the risk neutral probabilities of upward or downward movements of the underlying stochastic process. Doing this iteratively forward in time, the resulting graphical structure forms a recombining triangular tree for the time and state space of the stochastic variable, e.g., the stock price. As time intervals get arbitrarily small the original underlying stochastic process is recovered in the limit. Once having built the triangular tree it is solved backwards step by step in a dynamic programming fashion. At each node in the tree the corresponding option value is determined along with the optimal exercise strategy. For example, for an American put option at each node it is checked whether the option value alive is larger than its intrinsic value or not. Finally, this procedure yields the value of the option at time zero depending on the current realization of the underlying and the optimal exercise strategy, i.e., the values of the underlying that trigger exercise. This simple example of an American option can be extended to more complex real option interactions with different sources of uncertainty (including mixed jump-diffusion processes and multidimensional processes), dividend-like payments, and reversible as well as irreversible European or American style switches. Furthermore, multinomial triangular lattices can be constructed that are more accurate than the simple binomial tree that only considers one upward or downward step of the underlying source of uncertainty per time step.[19]

Triangular lattice techniques have several advantages. In particular, they are:

- easy to understand,

- computationally easy to implement, and

- provide guidelines for optimal managerial decisions in response to market conditions.

However, there exist some serious drawbacks of triangular lattice techniques which are especially important for the valuation of complex real option interactions. Generally, there

- exists a tradeoff between accuracy and time consumption depending on the number of time steps of the time partition,

- might occur convergence problems,

- exist problems in the presence of path dependencies,

- might be simple decision rules resulting from the determination of the continuation regions of generalized timing and switching options which will not be detected, and

---

[19]For an in-depth treatment of lattice techniques in real option valuation see, e.g., Ritchken/Kamrad [191], Kamrad [106], Trigeorgis [222] and [225], chapter 10, Calistrate/Paulhus/Sick [40], Koch [115], chapter 3, and Bollen [28, 29] for a pentanomial tree to model the product life cycle considered here.

- might occur problems concerning the smoothness of the resulting value function.

The last topic is especially cumbersome for the stochastic control approach taken in this study because the verification theorems for the generalized timing and switching options are not applicable to triangular lattice techniques directly. Therefore, we cannot be sure that the value function obtained by solving the lattice indeed converges to the optimal solution of the corresponding combination of impulse control and optimal stopping problems.

The general idea of Monte Carlo simulation is to construct a triangular lattice over time and states for which, starting from an initial point, many paths of the underlying are simulated and the resulting value of the option is determined. The approximate market value of the option can be calculated by averaging over all simulated outcomes. For example, in order to value a European call using Monte Carlo simulation, many price paths starting from the current value of the underlying are drawn randomly using the risk neutral probability measure up to the exercise date of the option. Averaging over all obtained outcomes of the underlying and the option´s expiration date and discounting at the riskless interest rate yields an approximation of the true market value of the option. The main drawback, which makes Monte Carlo simulation not feasible for valuing real options, is that only European options can be valued because the lattice is folded forward in time. However, Monte Carlo simulation is a very powerful tool once we have determined the optimal exercise strategies and continuation regions of the interacting real options. Using the optimal exercise strategy we can simulate trajectories of the underlying sources of uncertainty along with the resulting project values. This results for each initial starting point in the distribution function of the project value of the option interaction model. Determining the risk profile of the project value in this way is crucial to further analyze the risk exposure of the firm if the particular project with its associated collection of real options is realized. Risk profiles for each of the above product life cycle models resulting from simulations are presented in Section 6.5.

6.3.1 Finite Difference Methods

Since in the stochastic control approach to real option pricing all partial differential equations for the incorporated option values are given in terms of systems of quasi-variational inequalities that a numerical technique that approximates the PDEs directly is preferable to the above mentioned numerical methods. The main representatives with regard to approximating the solution of a system of PDEs are the *finite difference methods*. With finite difference methods the PDEs describing the evolution of the option values are approximated by a set of difference equations, which are then solved in a backward recursive fashion taking care of the boundary conditions the value functions satisfy.

In order to illustrate the approach we consider how it might be used to derive the project value in the decay regime with flexible capacity $V_{Fd}(t, \Theta_t, M_t)$ and the corresponding optimal capacity expansion and contraction policy. The PDE the project value must satisfy for all $X_t = (t, \Theta_t, M_t)$ in the continuation region C_d is

$$\frac{1}{2}\sigma^2\Theta^2\frac{\partial^2 V_{Fd}}{\partial\Theta^2} + \mu_d\Theta\frac{\partial V_{Fd}}{\partial\Theta} + \frac{\partial V_{Fd}}{\partial t} - rV_{Fd} + f = 0. \tag{6.69}$$

The procedure now starts by creating a rectangular discretization of the state space with regard to time, the underlying sources of uncertainty and the action space. A number of equally spaced times between time zero and the end of the product life cycle T are chosen. Suppose that $\Delta t = T/N_t$ and consider the $N_t + 1$ times

$$0, \Delta t, 2\Delta t, \ldots, T. \tag{6.70}$$

A number of equally spaced demand parameter levels and capacity levels are also chosen. Suppose that Θ_{max} is high enough such that it may almost never be reached from the current demand parameter level Θ_0. We define $\Delta\Theta = \Theta_{max}/N_\Theta$ and consider a total of $N_\Theta + 1$ demand parameter levels

$$0, \Delta\Theta, 2\Delta\Theta, \ldots, \Theta_{max}. \tag{6.71}$$

The same discretization is performed for the possible capacity levels. Define $\Delta M = (M_{max} - M_{min})/N_M$ and consider the $N_M + 1$ capacity levels

$$M_{min}, M_{min} + \Delta M, M_{min} + 2\Delta M, \ldots, M_{max}. \tag{6.72}$$

The resulting grid consists of $(N_t + 1)(N_\Theta + 1)(N_M + 1)$ points. The (i, j, k) point on the grid is the point that corresponds to time $i\Delta t$, demand parameter level $j\Delta\Theta$ and capacity level $k\Delta M + M_{min}$. We will use the variable $V_{i,j,k}$ to denote the project value with flexible capacity in the decay regime at the point (i, j, k).

In general, there are now several ways in which the partial derivatives of the PDEs can be approximated by finite differences. Since we are interested in solving many PDEs in parallel due to the $N_M + 1$ different capacity levels, we will concentrate solely on the *explicit finite difference scheme* because of its computational efficiency. Although other difference approximations, like implicit finite differences or the Crank–Nicholson scheme,[20] may be more robust, using the explicit scheme consumes the least computer time. Furthermore, stability and convergence problems that may occur can be overcome if the method is properly applied. In the explicit finite difference scheme the partial derivatives at any interior point (i, j, k) in the grid are approximated as[21]

$$\frac{\partial V}{\partial t} \approx \frac{V_{i+1,j,k} - V_{i,j,k}}{\Delta t} \tag{6.73}$$

$$\frac{\partial V}{\partial\Theta} \approx \frac{V_{i+1,j+1,k} - V_{i+1,j-1,k}}{2\,\Delta\Theta} \tag{6.74}$$

$$\frac{\partial^2 V}{\partial\Theta^2} \approx \frac{V_{i+1,j+1,k} - 2V_{i+1,j,k} + V_{i+1,j-1,k}}{\Delta\Theta^2} \tag{6.75}$$

---

[20] See Hull [102], p. 369, or Trigeorgis [225], p. 313.
[21] See Trigeorgis [225], p. 315.

where the partial derivatives with respect to Θ are approximated at point $(i+1, j, k)$ instead of (i, j, k). That means it is implicitly assumed that the partial derivatives with respect to Θ one time step ahead at $i+1$ are the same as at the current time i. Substituting the difference approximations into the PDE (6.69) yields

$$\frac{1}{2}\sigma^2 (j\,\Delta\Theta)^2 \frac{V_{i+1,j+1,k} - 2V_{i+1,j,k} + V_{i+1,j-1,k}}{\Delta\Theta^2} \tag{6.76}$$
$$+ \mu_d\, j\,\Delta\Theta \frac{V_{i+1,j+1,k} - V_{i+1,j-1,k}}{2\,\Delta\Theta}$$
$$+ \frac{V_{i+1,j,k} - V_{i,j,k}}{\Delta t} - r V_{i,j,k} + f_{i,j,k} \;=\; 0$$

which can be solved for

$$V_{i,j,k} = \frac{p_j^+\, V_{i+1,j+1,k} + p_j^0\, V_{i+1,j,k} + p_j^-\, V_{i+1,j-1,k} + f_{i,j,k}\,\Delta t}{1 - r\,\Delta t} \tag{6.77}$$

where

$$p_j^+ = \frac{1}{2}\sigma^2\, j^2\,\Delta t + \frac{1}{2}\mu_d\, j\,\Delta t, \tag{6.78}$$

$$p_j^0 = 1 - \frac{1}{2}\sigma^2\, j^2\,\Delta t, \tag{6.79}$$

$$p_j^- = \frac{1}{2}\sigma^2\, j^2\,\Delta t - \frac{1}{2}\mu_d\, j\,\Delta t, \tag{6.80}$$

for $i = 0, \ldots, N_t - 1$, $j = 1, \ldots, N_\Theta$ and $k = 0, \ldots, N_M$. Since we calculate project values backward in time starting at the terminal time T, the project values one time step ahead at $i+1$ are known at time step i. Therefore, the only unknown in equation (6.77) is $V_{i,j,k}$ which can be calculated explicitly. Hence, the name explicit finite difference scheme. Note that the weights p_j^+, p_j^0 and p_j^- can be interpreted as the risk neutral probabilities of a trinomial tree.[22] As probabilities they should be nonnegative and add up to 1. However, for large j these conditions may be violated resulting in instability problems.

For that reason Brennan/Schwartz [33] proposed to use the finite difference method with $D = \log \Theta$ rather than Θ as the underlying variable. Performing the corresponding transformation the PDE (6.69) becomes

$$\frac{1}{2}\sigma^2 \frac{\partial^2 V_{Fd}}{\partial D^2} + \left(\mu_d - \frac{\sigma^2}{2}\right) \frac{\partial V_{Fd}}{\partial D} + \frac{\partial V_{Fd}}{\partial t} - r V_{Fd} + f = 0. \tag{6.81}$$

The grid evaluates the derivatives for equally spaced values of D rather than for equally spaced values of Θ. The difference equation for the explicit method is then

---

[22]See Trigeorgis [225], p. 315.

given by

$$\frac{1}{2}\sigma^2 \frac{V_{i+1,j+1,k} - 2V_{i+1,j,k} + V_{i+1,j-1,k}}{\Delta D^2} \tag{6.82}$$

$$+ \left(\mu_d - \frac{\sigma^2}{2}\right) \frac{V_{i+1,j+1,k} - V_{i+1,j-1,k}}{2\,\Delta D}$$

$$+ \frac{V_{i+1,j,k} - V_{i,j,k}}{\Delta t} - rV_{i,j,k} + f_{i,j,k} = 0$$

or

$$V_{i,j,k} = \frac{p^+ \, V_{i+1,j+1,k} + p^0 \, V_{i+1,j,k} + p^- \, V_{i+1,j-1,k} + f_{i,j,k}\,\Delta t}{1 - r\,\Delta t} \tag{6.83}$$

where

$$p^+ = \frac{\Delta t}{2\,\Delta D^2}\sigma^2 + \frac{\Delta t}{2\,\Delta D}\left(\mu_d - \frac{\sigma^2}{2}\right), \tag{6.84}$$

$$p^0 = 1 - \frac{\Delta t}{\Delta D^2}\sigma^2, \tag{6.85}$$

$$p^- = \frac{\Delta t}{2\,\Delta D^2}\sigma^2 - \frac{\Delta t}{2\,\Delta D}\left(\mu_d - \frac{\sigma^2}{2}\right). \tag{6.86}$$

For the log-transformed problem the weights p^+, p^0 and p^- are independent of j which prevents stability problems occurring for the untransformed problem. Again the weights can be interpreted as probabilities of upward, sideward or downward changes of the demand parameter level Θ in the next small time interval Δt to the new values $\Theta e^{\Delta D}$, Θ, and $\Theta e^{-\Delta D}$ respectively. Choosing the time step Δt and the log-demand parameter increment ΔD such that

$$\Delta t \leq \frac{\sigma^2}{(\mu_d - \frac{1}{2}\sigma^2)^2}, \tag{6.87}$$

$$\Delta D \leq \frac{\sigma^2}{|\mu_d - \frac{1}{2}\sigma^2|}, \tag{6.88}$$

the weights are nonnegative and stability is ensured. As Hull [102] points out, the log-transformed explicit finite difference method attains its highest numerical efficiency for $\Delta D = \sigma\sqrt{3\,\Delta t}$.

In order to calculate all values $V_{i,j,k}$ backwards for $i = N_t - 1, \ldots, 0$ we have to impose three kinds of boundary conditions. The first takes care of the fact that, by assumption, the project is worthless at time $T = N_t\,\Delta t$. It is given by the terminal condition (6.49)

$$V_{N_t,j,k} = -c_6\,(M_{min} + k\,\Delta M) - c_7 \qquad \forall j, k. \tag{6.89}$$

Next we need to consider the upper and lower boundaries of the grid with respect to Θ. If j is zero, further downward changes are precluded. Therefore an extrapolation of the value function has to be chosen in order to find the value of the preceding time step using equation (6.83). The same situation occurs at the upper boundary for $j = N_\Theta$. Several approaches are conceivable to perform such an extrapolation. We chose to extrapolate the value function linearly downwards or upwards, respectively. Although this choice is somewhat arbitrary the numerical results in the next section suggest that the numerical procedure is quite robust with respect to changes in the kind of extrapolation, as long as the number of considered demand parameter levels is sufficiently large and we are interested in project values in the middle range of Θ.

After having calculated the project values at time i for each $j \in \{0, \dots, N_\Theta\}$ and $k \in \{0, \dots, N_M\}$ from the subsequent given project values at time $i + 1$, the last kind of boundary condition comes into play. For each combination of demand parameter level j and capacity level k it has to be determined whether switching to another capacity level k' is optimal or not. According to the maximum operator defined by equation (6.43) switching is optimal if

$$V_{i,j,k} < \max_{k' \neq k}\{V_{i,j,k'} - H_{E/C}(i, j, k, k')\}. \tag{6.90}$$

In case switching to the new capacity k' is optimal, the new project value becomes

$$V_{i,j,k} = V_{i,j,k'} - H_{E/C}(i, j, k, k'). \tag{6.91}$$

This yields for each combination (i, j, k) the new optimal capacity level k' regardless of whether (i, j, k) is already in the continuation region or not. If $k' = k$ there is no change in capacity and the system is still in the continuation region C_d. If $k' > k$, then expansion is optimal and the upper boundary of the continuation region at time i can be found by selecting for each k the demand parameter level $j^*(i, k)$, i.e., $\Theta_d^*(t, M_t)$, for which k' is the first time greater than k. Then set $k^*(i, k) = k'$, i.e., $M_d^*(t, M_t)$, as the corresponding new capacity level to which production is expanded.[23] Similarly the lower boundary of the continuation region can be found by determining the demand parameter level $j^{**}(i, k)$, i.e., $\Theta_d^{**}(t, M_t)$, for which k' is the first time smaller than k. If the demand parameter level hits the boundary $j^{**}(i, k)$ at time i and current capacity k from above, then the capacity contraction to the new level $k^{**}(i, k)$, i.e., $M_d^{**}(t, M_t)$, is performed. Therefore the free boundary condition (6.90) implicitly determines the form of the continuation region for each capacity level k at time i. Working backward in time we finally get the project values at time zero $V_{0,j,k}$ for each initial demand parameter level j and initial capacity k along with the optimal switching strategy for each point (i, j, k) in the grid.

---

[23] Due to the triangular inequality that is satisfied by the switching cost function $H_{E/C}$ the new grid point (i, j, k') must be in the continuation region. Consequently, at each point in time i, at maximum one capacity expansion or contraction takes place.

Note that the optimal expansion and contraction policy is determined for each point inside the intervention region as well, i.e., for all points in the grid where either $j > j^*(i,k)$ or $j < j^{**}(i,k)$. If the production system starts at time 0 in the continuation region, these points in the grid can never be attained at time i because the process would have been forced back into the continuation region before. However, they are necessary to calculate the project values at time $i-1$. Furthermore, it is worth noting that the specific form of the capacity expansion and contraction policy determined by the continuation region is automatically received in the numerical procedure and results in an impulse control strategy. Assuming convergence of the explicit finite difference scheme we are therefore able to conjecture that the numerical solution of the sufficient quasi-variational inequalities is indeed the solution of the corresponding impulse control problem.

The same procedure can be used to numerically determine the other project and option values and their optimal exercise strategies incorporated in the product life cycle model. For the timing and intensity option in the decay regime with flexible capacity, for example, the option value $V_{WFd}(t, \Theta_t, 0)$ satisfies the PDE

$$\frac{1}{2}\sigma^2\Theta^2\frac{\partial^2 V_{WFd}}{\partial\Theta^2} + \mu_d\Theta\frac{\partial V_{WFd}}{\partial\Theta} + \frac{\partial V_{WFd}}{\partial t} - rV_{WFd} = 0 \qquad (6.92)$$

in the continuation region C_{Fd}^ϑ. This translates into the following explicit finite difference scheme at time i and demand parameter level j where $V_{WFd}(t, \Theta_t, 0) = W_{i,j}$:

$$W_{i,j} = \frac{p^+ W_{i+1,j+1} + p^0 W_{i+1,j} + p^- W_{i+1,j-1}}{1 - r\,\Delta t} \qquad (6.93)$$

and $k = 0$ is omitted. The boundary conditions are similar as above. The terminal condition is $W_{N_t,j} = 0$ for all demand parameter levels j. The upper and lower boundaries with respect to Θ are determined by linear extrapolation. The only difference from the flexible capacity option is that at each time i only one free boundary condition instead of N_M has to be considered for each demand parameter level j. From the maximum operator (6.59) follows that we have to check for each j at time i if it is optimal to start operation with initial capacity k' or stay idle and wait to invest. This means that if

$$W_{i,j} < \max_{k'\neq0}\{V_{i,j,k'} - H_I(i, j, k')\}, \qquad (6.94)$$

starting operation with the initial capacity k' is optimal and the new option value becomes

$$W_{i,j} = V_{i,j,k'} - H_I(i, j, k'). \qquad (6.95)$$

Otherwise the time demand combination (i, j) is in the continuation region and the firm remains waiting. Furthermore, the smallest demand parameter level $j_{Fd}^\vartheta(i)$ for which the inequality (6.94) holds determines the boundary of the continuation region. If it is hit from below the firm starts operations at time i with initial optimal capacity $k_{Fd}^\vartheta(i) = k'$.

6.3.2   General Numerical Solution Procedure

After having demonstrated the main ideas of finite difference methods and how they can be used for the valuation of the product life cycle model, we will summarize the main steps of the numerical solution procedure starting from the graphical decomposition of the contingency structure. From the graphical decomposition all information concerning the incorporated states, their earliest entry and latest exit time, the nature of reversible and irreversible switches, the underlying sources of uncertainty in each state along with the action spaces as well as the cash flow and switching functions are known. Using this information the following general procedure to transform the graphical decomposition structure into a numerical valuation scheme using finite difference methods is proposed:

1. Determine the systems of quasi-variational inequalities for each generalized switching option.

2. Determine the set of variational inequalities for each generalized stopping option.

3. Discretize appropriately over time, sources of uncertainty and action space.

4. Determine finite difference approximation of the PDEs in each state for the time intervals the states are valid.

5. Determine the three different kinds of boundary conditions (terminal or initial conditions, upper and lower boundaries of the sources of uncertainty, free boundary conditions for switching) for each state.

6. Go one step backwards in time and calculate the project values in each state.

7. Check for each possible realization of the sources of uncertainty whether switching to another state is possible and optimal.

8. Store the optimal switching policy for each state.

9. Go back to step 5 until time zero is reached.

This algorithm yields the time zero value of the project in each state for all possible initial values of the underlying stochastic processes. Furthermore, the optimal exercise or switching policies are derived for each time, state and realization of the underlying uncertainties over the whole lifetime of the project.

In view of the product life cycle model above the described valuation procedure is self explanatory. However, there are some important cases which are not covered in the product life cycle model. There it was assumed that the earliest entry (time zero) and latest exit time (time T) for all states were identical. However, this need not be the case for other applications. Therefore it is important to determine the time intervals for each irreversible switch where switching to another state is possible or not. This may give rise to initial boundary conditions in addition to terminal conditions. Furthermore, in the product life cycle model the firm is in all states (waiting and operating in the growth or decay regime, respectively) exposed to the same two sources of uncertainty, namely demand risk and regime switching risk. Again other applications might not admit this simplifying feature. For example, consider the decision to switch production from a domestic plant to a production facility abroad. Assuming that production is completely sold at home, the state producing at the foreign plant is exposed to exchange rate uncertainty while the state producing at the domestic plant is not. In order to take the foreign investment decision into account we need to model *all* sources of uncertainty over the whole project life in parallel, although the opportunity to produce abroad may not be available now but only in the future.[24] While many different sources of uncertainty cause no principle problem for the numerical valuation procedure, multidimensional finite difference schemes are yet to be employed for the solution of real option interaction models.[25]

6.4   Numerical Analysis

In order to understand the basic properties of the model, we start our discussion of the numerical results with the optimal instantaneous production rate Q^* and the corresponding form of the instantaneous cash flow function f, as well as the price function for one unit of output if markets are cleared immediately. Afterwards we will compare the value functions and the optimal initial capacity choice decisions for all four models of Table 6.1. Simulation results will be presented which gives further rise to risk profiles of each of the four models. The risk profiles can be used to show how flexibility reduces the downside risk of an investment while fully exploiting the upside potential of the uncertain product life cycle. As it will turn out, flexibility with respect to capacity changes is extremely valuable. In order to get a detailed feeling for the complex nature of the impulse control problem underlying the flexible capacity option, a typical trajectory for the demand parameter process is picked that clarifies the basic properties of capacity expansion and contraction. Furthermore, we will discuss the impact of capacity restrictions and the role of the option to abandon on the project values and the optimal operating policies. The robustness of the results with respect to changes in different parameter values will

---

[24]Of course, the underlying sources of uncertainty need to be observable today regardless of whether they affect a state whose first possible entry time is well in the future or not. In the example of exchange rates there should be no problem with observability. However, in other cases where the strategy space of the firm is itself uncertain these observability problems may arise. See also the discussion in Section 2.2 about strategy space risk.

[25]To the current date real option applications with more than two underlying sources of uncertainty modelled as a diffusion process have not been presented in the literature. Although some authors discuss general solution procedures for multidimensional problems, see e.g., Hodder/Triantis [96], numerical results are still missing. However, it should be one of the main tasks of real option research in the future to find numerical solution techniques which work efficiently in a multidimensional setting. One approach could be to employ the multidimensional extensions of finite difference methods that are well known in modern physics, see e.g., Großmann/Roos [86].

be analyzed. Finally, the economic implications as well as the implications for real option interaction models will be addressed.

Parameter	Value	
Initial demand parameter	$\Theta_0=1.0$	units
Demand function slope	$\lambda=1.0$	units per DM
Riskless interest rate	$r=0.05$	annually
Risk neutral drift growth regime	$\mu_g=0.30$	annually
Risk neutral drift decay regime	$\mu_d=-0.30$	annually
Demand parameter volatility	$\sigma=0.20$	annually
Expected regime switching time	$\mu_\tau=5.0$	years
Regime switching standard deviation	$\sigma_\tau=1.0$	years
Time horizon	$T=10.0$	years
Linear production cost parameter	$c_1=0.4$	DM per year and unit produced
Quadratic production cost parameter	$c_2=0.5$	DM per year and square unit produced
Overhead cost parameter	$c_3=0.3$	DM per year and unit installed capacity
Investment cost	$c_4=2.0$	DM per unit capacity
Variable expansion cost	$c_5=2.0$	DM per unit capacity
Variable contraction cost	$c_6=0.4$	DM per unit capacity
Fixed cost of capacity change	$c_7=0.1$	DM

Table 6.2: Base case parameter values

In order to find numerical solutions of the four above described models of production flexibility we use the base case parameters displayed in Table 6.2. The product life cycle starts at an initial demand parameter level of $\Theta_0 = 1.0$ and evolves stochastically with drift rate of $\mu_g = 0.3$ up to the normally distributed regime switching time τ where it enters the decay regime with drift $\mu_d = -0.3$. The product life cycle is assumed to last $T = 10$ years. The volatility σ is assumed to be equal in the growth and in the decay regime. The expected regime switching time τ is set such that the product life cycle is on average symmetric. For simplicity the unit expansion cost c_5 is set equal to the initial unit investment cost c_4. Furthermore, it is assumed that unnecessary capacity can be sold for 20% of its initial value, i.e., $c_6 = 0.2c_4 = 0.4$. The fixed cost incurred at each change of capacity after the initial investment is $c_7 = 0.1$.

For the explicit finite difference scheme the time, demand parameter and capacity grid is constructed in the following way. Δt was set to 0.01 resulting in $N_t = 1000$ time steps over the whole product life cycle of ten years. For the demand parameter dimension of the grid a log-transformed step size of $\Delta D = 0.03$ is chosen which corresponds approximately to $\sigma\sqrt{3\Delta t}$. The number of considered demand parameter levels is $N_\Theta = 200$ starting form $\Theta_0 = 1.0$ with 100 upward and downward steps, respectively. Finally, the discretization with respect to the capacity levels uses a step size of $\Delta M = 0.05$ with $N_M = 200$ starting at $M_{min} = 0$ up to $M_{max} = 10.0$. An appropriate choice of the step sizes and overall size of the grid dimensions are crucial for the numerical accuracy of the finite difference scheme. We performed several runs of the computer program in order to ensure stability of the procedure. It turned out that a further decrease in Δt does not change the results significantly (less than 0.1%). The number of capacity levels N_Θ is the parameter choice that affects the valuation results most seriously. Because the probability of reaching the highest or lowest demand level in the grid starting in the midst at Θ_0 is practically zero with $N_\Theta > 140$, the results are quite stable for sufficiently large N_Θ at least for the $N_\Theta/2$ values in the middle of the grid which we are primarily interested in. Furthermore, our numerical results suggest that for $\Delta M < 0.1$ the numerical solution procedure rapidly converges. For the selected grid size the explicit finite difference scheme for the four models was implemented in a C++ program on a Pentium 166 computer. The computer time consumed to solve for the value functions of the four models in the growth and the decay regime along with their corresponding optimal exercise strategies lies between 60 seconds and 7 minutes, showing that the method is computationally quite efficient.

We now start with the numerical analysis of the model for the base case parameters. Figures 6.5 to 6.7 show the impact of changes in the demand parameter level Θ on the optimal production rate Q^*, the instantaneous cash flow f and the market price for one unit of output P for the capacity levels $M = 1.0$ and $M = 1.5$, respectively. Starting with the optimal production rate of equation (6.16) the production rate is zero if $\Theta < \lambda c_1 = 0.4$ independent of the currently installed capacity M. On the other hand the firm operates at full capacity if $\Theta > 2M + \lambda(c_1 + c_2)$. For $M = 1.0$ this upper limit is $\Theta = 2.9$ and for $M = 1.5$, $\Theta = 3.9$, respectively. The optimal instantaneous production rates for the base case parameters are displayed in Figure 6.5. As can be seen from the figure the optimal production rate is linear in Θ for intermediate values. The ability to control production rates in a range from zero to M actually includes the state of costly suspension for values of $\Theta < 0.4$.

Depending on the currently installed capacity M, suspended operation causes

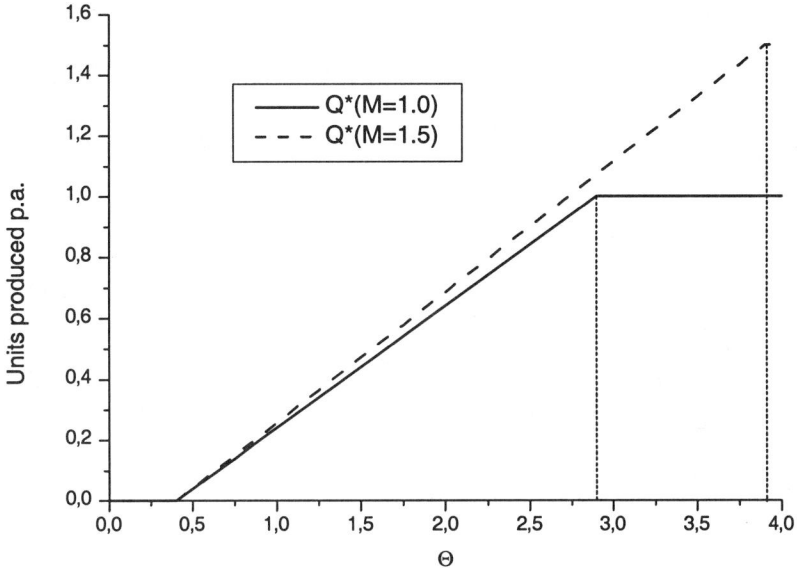

Figure 6.5: Optimal production rate Q^* for capacity levels $M = 1.0$ and $M = 1.5$

instantaneous overhead cost of $c_3 M = 0.3M$. Therefore the instantaneous cash flow f is negative with zero production rate and the lower the rate, the higher the currently installed capacity. For the assumed initial demand parameter level $\Theta_0 = 1.0$ the cash flow function is still negative for both displayed capacity levels $M = 1.0$ and $M = 1.5$ as illustrated in Figure 6.6. As Θ rises the higher overhead cost of the capacity level $M = 1.5$ in comparison to $M = 1.0$ is offset by the ability to produce more, which results in larger instantaneous cash flows. Therefore it is obvious that the option to adjust capacity adds value to the project. Not being committed to a specific capacity level over the whole lifetime of the project can have significant advantages. If demand drops it is favorable to have less capacity installed because overhead costs are lower. On the other hand if demand increases the firm may want to exploit the higher potential cash flow from operating at a higher capacity. For example for demand parameter levels higher than 3.3, the instantaneous cash flow of 1.5 units installed capacity exceeds the cash flow earned by a production facility with one unit of capacity.

The output price of one produced unit fulfills the relationship (6.11). If markets are cleared instantaneously the price per unit output for the optimal production rate Q^* is displayed in Figure 6.7 as a function of Θ. In the range of costly suspension

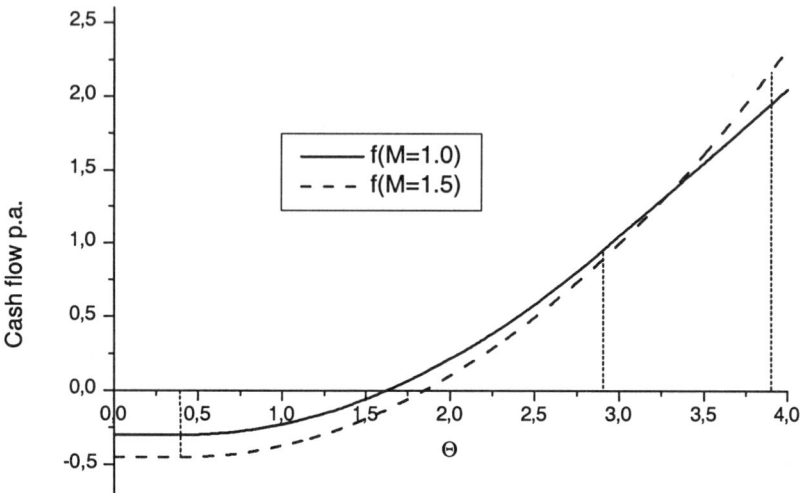

Figure 6.6: Instantaneous annual cash flow f for capacity levels $M = 1.0$ and $M = 1.5$ with optimally controlled production rates

	Immediate Investment	Timing and Intensity Option
Static Project	$V_{Pg} - c_4 M_0 = 4.057$ $(M_0 = 1.10)$	$V_{Wg} = 5.740$ $(\Theta_g^\vartheta = 4.482, M_g^\vartheta = 7.50)$
Flexible Capacity	$V_{Fg} - c_4 M_0 = 7.343$ $(M_0 = 0.30)$	$V_{WFg} = 7.427$ $(\Theta_{Fg}^\vartheta = 1.665, M_{Fg}^\vartheta = 0.60)$

Table 6.3: Project and option values for base case parameters

the firm produces nothing and the output price rises linearly with slope $1/\lambda = 1.0$. If the firm is operating at full capacity, i.e., if $\Theta > 2M + \lambda(c_1 + c_2)$, the slope of the price function is again $1/\lambda$. In both cases the firm keeps its output rate constant. Hence the only effect that changes the unit output price comes from changes in the demand parameter level. If however the firm adjusts its output rate to changes in demand, an increase in demand is partially offset by higher production rates leading to a lower increase in unit output prices. The slope of the price function

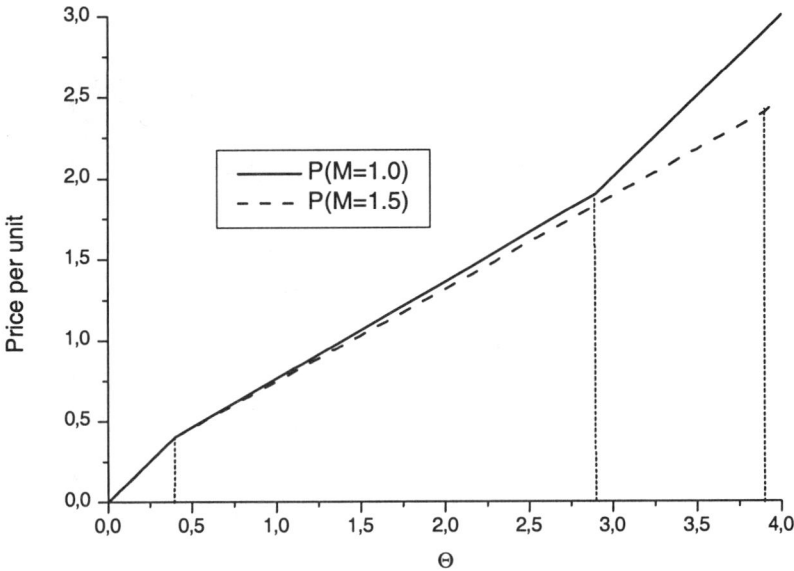

Figure 6.7: Output price per unit produced for optimally controlled production rates

for intermediate values of Θ can be obtained as

$$m = \frac{1 + \frac{\lambda c_2}{M}}{\lambda(2 + \frac{\lambda c_2}{M})}. \tag{6.96}$$

It can easily be verified that we have $1/(2\lambda) < m < 1/\lambda$, since c_2, λ and M are positive. Although the firm achieves a higher price at lower capacity it is not necessarily optimal to produce at low capacity. For example for high demand parameter levels the decrease in price by operating at higher capacity is overcompensated by an increase in output rate and a decrease in per unit overhead cost, making production at a higher capacity level more efficient. For the base case parameters at hand the value of the four models is displayed in Table 6.3. For the immediate investment case the optimal initial capacity to install, M_0, is given in brackets, while for both timing and intensity models the demand threshold at time zero Θ^ϑ along with the corresponding optimal capacity M^ϑ are written in brackets. Although the results cannot be generalized in a straightforward manner we discuss them in more detail.

First of all the static project value is already positive and a simple NPV rule would give the result to immediately build the plant with a capacity level of $M_0 =$

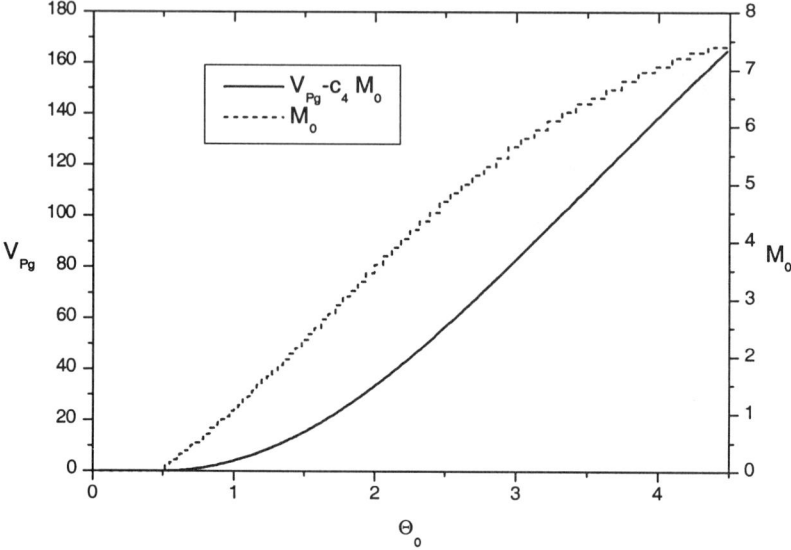

Figure 6.8: Static project value V_{Pg} and optimal initial capacity level M_0 as a function of initial demand parameter level Θ_0

1.10 units per year. Because the firm cannot adjust capacity during the product life cycle the initial capacity has to be chosen in such a way as to balance the large profits earned in the middle of the product life cycle against the losses at the beginning and at the end of the life cycle where demand is typically quite low.

Taking additionally the timing and intensity option into account, the situation significantly changes. The overall project value (static project value and option value) is 5.740 which reflects an increase of more than 41% of value. Obviously the timing and intensity option is extremely valuable for management. The time zero demand threshold $\Theta_g^{\vartheta} = 4.482$ is well above Θ_0, which makes postponing the investment optimal because the current demand parameter level is $\Theta_0 = 1.0$. Interestingly the initial optimal capacity level to install, $M_g^{\vartheta} = 7.50$, if the demand parameter level was $\Theta_0 = \Theta_g^{\vartheta} = 4.482$, is extremely high. The reason for this is that with a higher initial demand level the future demand parameter will reach higher values both in the growth and in the decay regime. Because capacity is fixed once the investment decision has been made, the firm is willing to install more capacity to reap the higher profits over the whole life cycle of the project. However, note that the optimal demand threshold and the optimal initial capacity

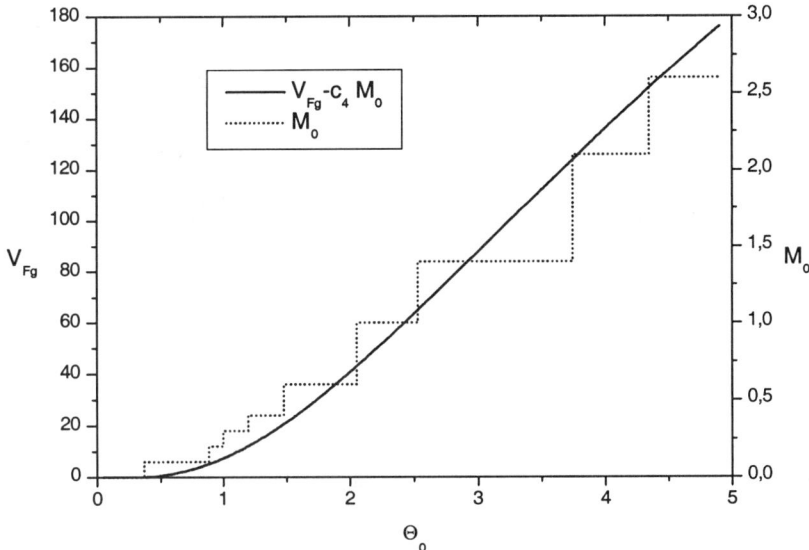

Figure 6.9: Flexible capacity project value V_{Fg} and optimal initial capacity level M_0 as a function of initial demand parameter level Θ_0

for the timing and intensity option depend on the time left in the product life cycle. We will elaborate on this point below.

The option to adjust capacity during the product life cycle is even more valuable than the timing and intensity option. In comparison to the static project value, flexible capacity adds 80% to the overall project value. As expected the initially installed capacity, $M_0 = 0.30$, is lower than for the static project because management has not to weigh future profits against current losses to the same extent as for the static project. Instead the firm will tend to start with low capacity and rapidly add capacity if demand rises before removing capacity if the product life cycle is in the decay regime and demand becomes sufficiently low.

Because the range of capacity levels for the flexible capacity option is assumed to be unlimited, i.e., capacity can be adjusted to arbitrarily high or low levels, the incremental value of the timing and intensity option is relatively small ($V_{WFg} - V_{Fg} = 7.427 - 7.343 = 0.084$). This is obvious because the flexible capacity option already contains the ability to delay investment of incremental units of capacity. The only source of additional flexibility in the timing and intensity option is that the firm is not committed to invest immediately. This is favorable for the base case parameters because adjusting capacity later incurs a fixed cost of $c_7 = 0.10$

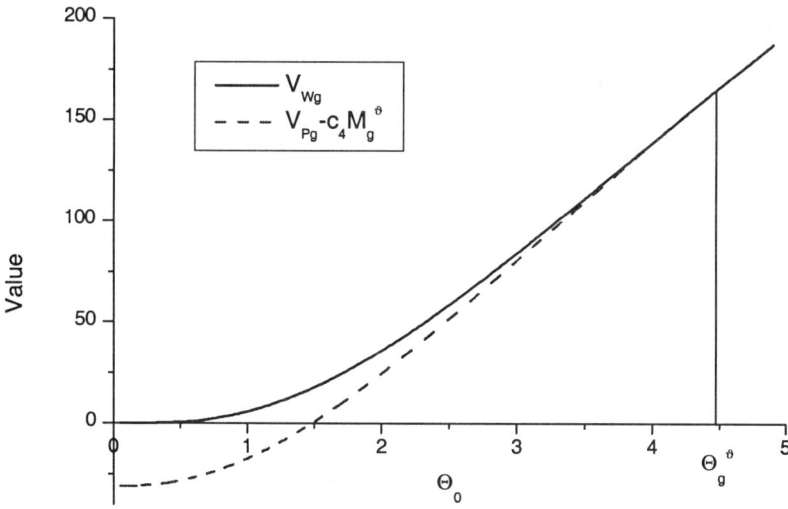

Figure 6.10: Timing and intensity option at time zero for static project with demand parameter threshold $\Theta_g^{\vartheta} = 4.482$ and optimal capacity $M_g^{\vartheta} = 7.50$

which has not to be paid for the initial investment. Together with the fact that the firm operating at $M_0 = 0.3$ right from the start already earns positive cash flows, the incremental value of the timing and intensity option is less than c_7. Furthermore, the demand threshold as well as the optimal capacity at time zero for which the timing and intensity option would be exercised immediately are much lower, $\Theta_{WFg}^{\vartheta} = 1.665$ and $M_{WFg}^{\vartheta} = 0.60$, than in the static project case. This is due to the fact that the ability to adjust capacity later makes the firm more willing to invest earlier in the project as soon as the cash flows are sufficiently positive.

An interesting aspect is how the four models are affected by changes of the initial demand parameter level Θ_0. Figures 6.8 and 6.9 show the impact of Θ_0 on the project value and the corresponding optimal capacity choice of the firm for the static project and the flexible capacity option, respectively. As expected both project value and initial capacity rise with higher initial demand. However, the differences in value between the static project and the flexible capacity project become negligible as soon as initial demand is very high. This is consistent with findings in the literature that real options are especially important for projects that are either out of the money or only slightly in the money.

The optimal capacity choice for the static project is an increasing function of initial demand. The kinks in the graph are due to the discretization of the capacity

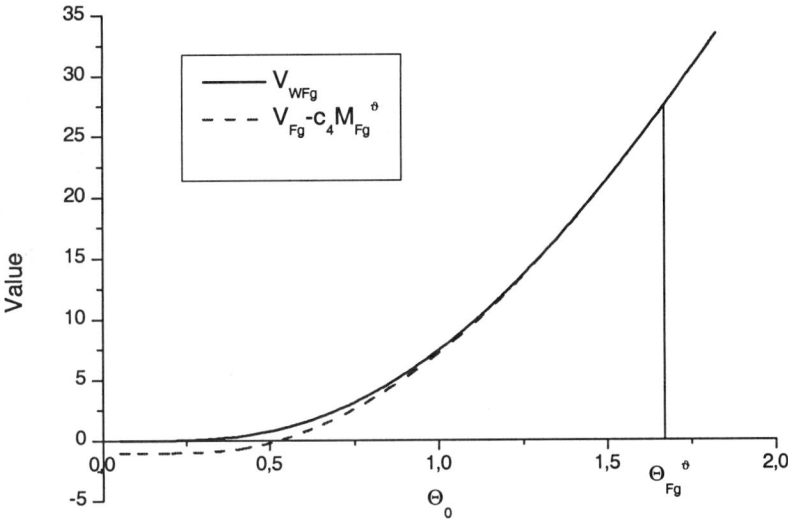

Figure 6.11: Timing and intensity option at time zero for flexible capacity project with demand parameter threshold $\Theta_{Fg}^{\vartheta} = 1.665$ and optimal capacity $M_{Fg}^{\vartheta} = 0.60$

levels. For the flexible capacity project the situation is somewhat different. The optimal initial capacity admits jumps larger than the capacity increment $\Delta M = 0.05$ because of the highly nonlinear structure of the infinitely many embedded expansion and contraction options of the model. Nevertheless as intuition suggests the optimal initial capacity choice is an increasing function of demand as well.

The value of the timing and intensity option for the static project and the flexible capacity project at time zero can be seen in Figures 6.10 and 6.11. From there it is obvious why the timing and intensity option adds less value in the presence of flexible capacity. The static project value for small initial demand parameter levels Θ_0 is very low, leading to a high difference to the value of the timing and intensity option which reflects the protective nature of the timing and intensity option for small demand levels. As can be seen from Figure 6.10 it takes relatively high demand parameter levels to compensate for this difference and to make exercising the option preferable at a demand threshold of $\Theta_g^{\vartheta} = 4.482$. On the other hand the protective nature of the timing and intensity option in the presence of flexible capacity, Figure 6.11, is rather limited because the difference in value is quite low even for small demand levels. Therefore the difference between the two value functions is offset for lower demand parameter levels leading to a demand threshold of $\Theta_{Fg}^{\vartheta} = 1.665$.

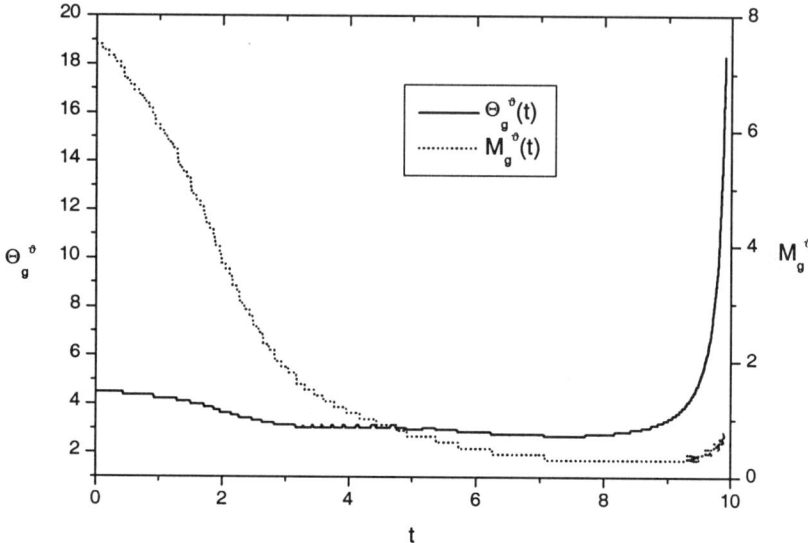

Figure 6.12: Demand parameter threshold Θ_g^{ϑ} and corresponding optimal capacity M_g^{ϑ} as a function of time t for the static project

For the timing and intensity option the time zero capacity choice is of rather limited importance, because at a given demand parameter level $\Theta_0 = 1.0$ the option is not exercised immediately. Since the investment threshold and the corresponding optimal capacity are now functions of time it is worthwhile to answer the question how they evolve over time. Figures 6.12 and 6.13 show these functional dependencies for the timing and intensity options on the static project and the flexible capacity project, respectively. Similar to the perpetual timing and intensity option discussed in Section 4.3, the exercise policy of the timing and intensity option for a product life cycle with finite time horizon is different from the pure timing option with fixed capacity. For example, for the static project the investment threshold Θ_g^{ϑ} is a decreasing function of time t up to year 7.5 and admits the above observed co-movement with the optimal capacity M_g^{ϑ}. After the seventh year the demand threshold sharply rises, reflecting the fact that the probability of further staying in the growth regime is almost zero and a drop of the demand parameter in the near future is most likely. Because the probability to be in the growth regime is very small after the eighth year, we state that in the most interesting time domain between time zero and year eight the investment threshold as well as the optimal

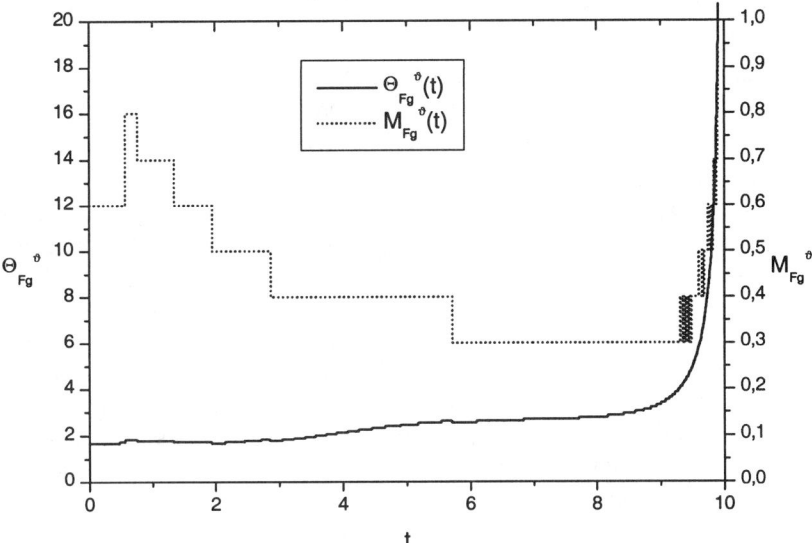

Figure 6.13: Demand parameter threshold Θ^{ϑ}_{Fg} and corresponding optimal capacity M^{ϑ}_{Fg} as a function of time t for the flexible capacity project

capacity level decrease. In comparison to that an otherwise equal option to wait with fixed capacity would show a strictly increasing investment threshold.[26] The answer to this phenomenon is simply given by the interplay of timing and intensity similar to what was already described in Section 4.3. At time zero the probability of regime switching in the near future is almost zero. In view of the high expected growth rate of the demand parameter, the firm will wait in order to invest only after the demand parameter has experienced a sufficiently high increase before it will commit itself to a capacity level for the rest of the product life cycle. However, as time goes by the threat of not being in the market when the regime switch occurs overcompensates for the value of waiting and the firm will invest at a lower demand level and will enter the market with less capacity than before. In other words, the opportunity cost of waiting increases leading to earlier investment. The situation for the pure timing option would lead to a strictly increasing demand threshold, because if the demand to be installed is fixed, the firm would be willing to invest earlier in expectation of a longer time to the end of the product life cycle. The logic behind this observation is the following. Since the firm cannot adjust capacity it

---

[26]See Dixit/Pindyck [72], p. 355.

will want to take advantage of the likely positive cash flows in the growth regime as long as possible, because it is not protected against the expected losses in the decay regime. Thus the investment threshold increases in time.

A similar logic applies to the behavior of the demand threshold of the timing and intensity option for the flexible capacity project. Now, the demand threshold Θ_{Fg}^{ϑ} slightly increases while the behavior of the optimal intensity of investment M_{Fg}^{ϑ} generally tends to decrease in the domain of interest. This is due to the ability to later revise capacity choice decisions. In light of later capacity adjustments the firm is willing to invest earlier with lower capacity in order to reap the benefits in early phases of the growth regime. This is caused by two contingencies. First, at low demand levels only production with low capacity earns positive cash flows. Therefore it is preferable to start with low capacity. Second, the flexibility to adjust production later makes the current investment decision less irreversible and hence raises the opportunity cost of waiting. Both conditions enhance investment. On the other hand after some time has passed and the firm has yet to invest, the expected time in the growth regime gets smaller and smaller. Thus, the probability of realizing high demand levels and high profits before the decay regime takes over is small. Therefore the opportunity cost of waiting decreases and investment is further delayed despite the ability to adjust capacity later. ˙

It is important to note that the similarity of the demand threshold behavior as a function of time for the timing and intensity option with flexible capacity and the pure timing option with fixed capacity is caused by completely different conditions. The former results from more flexibility in comparison to the timing and intensity option for the static project while the latter is caused by less flexibility.

In order to separate the effect of timing and intensity as well as flexible capacity at time zero, it is helpful to consider the question where the sources of value come from. In Figure 6.14 the part of value coming from the static project is compared to the option value added by the timing and intensity option. For low initial demand levels Θ_0 the option value is much larger than the original static project value. This effect is decreasing for larger demand levels and finally gets negligible as Θ_0 is well above 3.5 before it becomes zero at $\Theta_0 = \Theta_g^{\vartheta} = 4.482$. Again the downside protective nature of the timing and intensity option can be observed.

A similar but more pronounced effect is viewed for the flexible capacity option value in Figure 6.15. While the timing and intensity option only limits the downside risk for small initial demand parameter levels, the option to adjust capacity is also valuable for larger initial demand. Therefore the flexible capacity option is not only a means to reduce the downside risk by contracting capacity but to better exploit the upside potential of the product life cycle by expanding capacity. Although the value added by flexible capacity becomes negligible if initial demand gets large, we can conjecture from the figure that the flexible capacity option is generally preferable to the timing and intensity option as long as the possible capacity levels are not restricted to be in a certain range.

This finding becomes even more significant if all three sources of value are

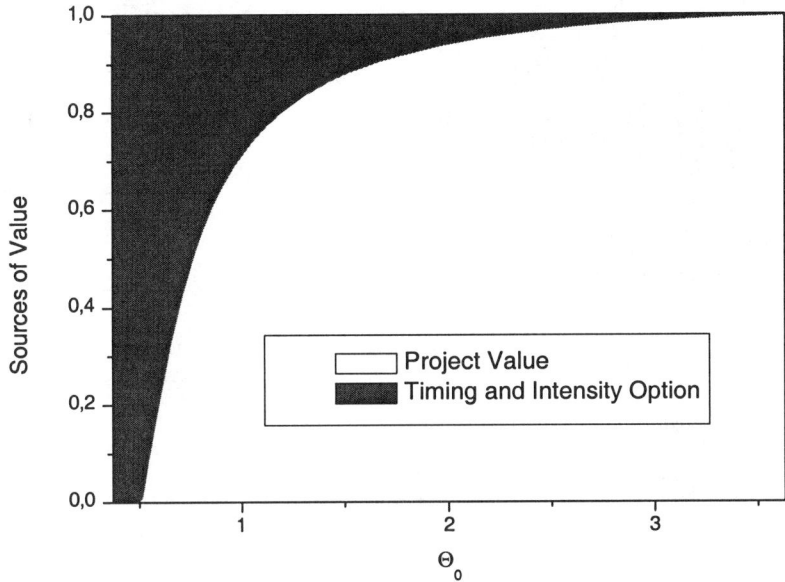

Figure 6.14: Normalized sources of value from the timing and intensity option and the static project value as a function of Θ_0

considered as in Figure 6.16. Since the timing and intensity option is now exercised at an initial demand parameter level of $\Theta_0 = 1.665$, it does not contribute any value for demand parameter levels beyond that threshold. On the other hand the protective nature of the timing and intensity option for low demand levels is of limited value in the presence of flexible capacity. The reason is that the flexible capacity option contains both options to wait to invest in incremental capacity and options to abandon incremental units of capacity, making the value added by an additional timing and intensity option very small.

This result has some important implications for the production strategies firms should pursue. Firms should rather exploit the flexibility inherent in their production systems than simply wait to invest until market conditions are favorable. Managing production flexibility is a source of significant strategic value which can outweigh the disadvantage of investing too early. Especially in the presence of product market competition, preemptive investment is often the only way to reap superior profits. While competition normally reduces the value part of the option to wait, it does even increase the share of option value of production flexibility, although this was not explicitly taken account of in our model. To be as early as

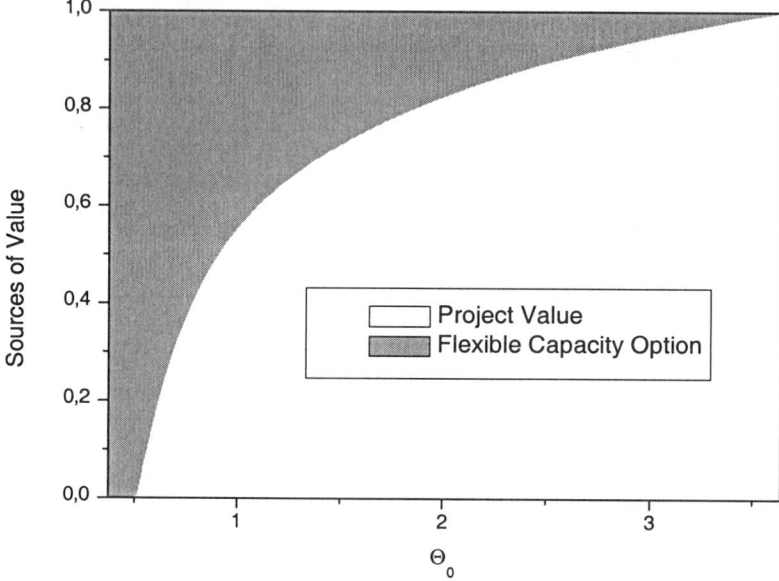

Figure 6.15: Normalized sources of value from the flexible capacity option and the static project value as a function of Θ_0

possible in the market and start production is often a competitive advantage which can be used only to its full extent if production flexibility is actively managed. Therefore the main result can be stated as follows. The optimal management of the flexibility inherent in a strategic asset might therefore be even more important than the selection of the optimal time to create the strategic asset; especially in the presence of competition.

Another important question to answer is how the value functions behave if input parameters change or are estimated wrongly. While the cost parameters of a flexible production system can be determined quite accurately, the estimation of the input parameters governing the exogenous sources of uncertainty are to be made with special care. There exist several problems with the volume flexibility models here. First of all the assumptions that overall demand is governed by a product life cycle model of the specific form proposed, and that the price demand relationship is a linear function have to be empirically checked. Provided that these assumptions hold the correct estimation of the corresponding input parameters is a crucial task. By performing a sensitivity analysis, the robustness of the valuation results with respect to misestimation of input parameters can be tested. Figure 6.17,

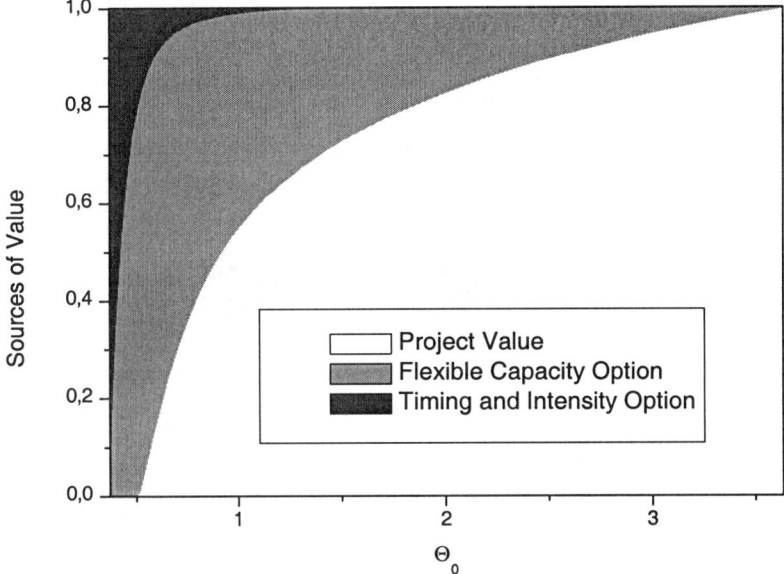

Figure 6.16: Normalized sources of value from the timing and intensity option, the flexible capacity option and the static project value as a function of Θ_0

for example, shows the changes in value of the four models if the corresponding input parameter on the left side is increased by 10% in comparison to the base case. From there it can be observed in which direction and to what extent input parameter changes affect the valuation results of all four models. We will shortly discuss the main results and their implications.

An increase in σ results in an increase in value which is the usually observed reaction of real option models on an increase in market priced risk. However, the impact of an increase in σ on the flexible capacity projects is much more pronounced than on the fixed capacity projects. This underlines the finding that flexible capacity makes the firm more capable of exploiting the upside potential of risk.

The changes of project values with an increase of the riskless interest rate r are as expected. Higher interest rates make projects less valuable. An increase in r has its largest impact on the value of the timing and intensity project. The reason for this is that the profits from the timing and intensity project come up only after several years due to postponing investment in the early phase of the product life cycle where the present value of cash flows is most valuable.

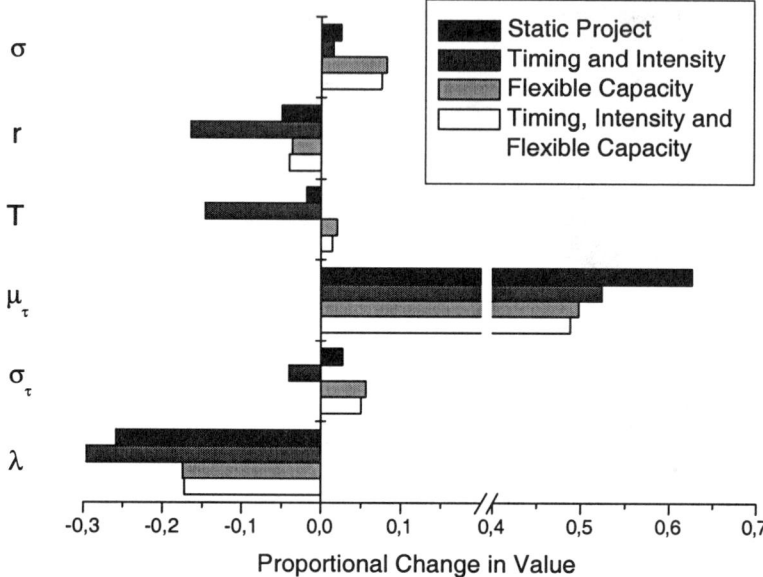

Figure 6.17: Impact of a 10% increase of input parameter on project values

An interesting statement can be made concerning the effect of an increase of the time horizon T. A larger time horizon means that the decay regime is expected to last longer than the growth regime. Therefore the probability of extremely low values of demand at the end of the product life cycle is high. Because the fixed capacity projects are committed to operate at a given capacity level until the end of the product life cycle, the negative cash flows incurred by relatively high capacity reduce overall project value. This effect is particularly marked for the timing and intensity project because installed capacity is higher than for the static project when exercise of the timing and intensity option is triggered. To the contrary, the ability to adjust capacity during the life cycle can even result in gaining value for longer time horizons. This is because the flexible capacity projects admit either abandonment if demand is extremely low or otherwise operating efficiently at low capacity levels.

The expected regime switching time μ_τ has the largest impact on project values of all input parameters. This becomes obvious from considering the two effects a higher μ_τ causes. First, the growth regime lasts longer resulting in significantly larger cash flows earned in the middle of the product life cycle. Second, regime switching takes place at larger demand parameter levels and the subsequent decay regime is shorter which leads to an upward shifted demand curve in the decay regime. Both effects admit earning positive cash flows longer for all four models. Although the valuation results are extremely sensitive with respect to the expected regime switching time μ_τ, the good news is that μ_τ is one of the input parameters that can be estimated quite accurately from past industry data.

Increasing the regime switching standard deviation σ_τ reveals another interesting result. While the value of the timing and intensity strategy with fixed capacity decreases with increasing σ_τ, all other value functions increase. A higher σ_τ increases the variety of possible product life cycles. While regime switching risk is not market priced, the value of the option to invest consequently decreases. For the timing and intensity option with flexible capacity the same effect can be observed, but it is offset by the ability to adjust capacity in the course of the remaining life cycle. It is important to note that changes in project values due to an increase in regime switching deviation are much less pronounced than changes due to an increase in expected switching time. For implementation purposes the estimation procedure should therefore lay more emphasis on the accuracy of the expected regime switching time than on the regime switching deviation.

Finally, the demand function slope λ is the second crucial input parameter. An increase in λ leads to a significant reduction in the value functions of all four models because prices are reduced more heavily by increasing output rates of the firm, making the investment less profitable. Estimating λ induces several difficulties. Among others the demand function slope neither needs to be constant in time nor be the same in the growth and in the decay regime. Moreover, the limited observability of the demand parameter Θ_t itself makes estimating λ an even more difficult task. Nevertheless, the existence of market reports for many industries and additional field studies of marketing divisions may help in this regard.

In summary, we find that the sensitivity of the project values with respect to different input parameters describing the environment of the firm is significant. Especially the value of the timing and intensity project with fixed capacity is not very robust to changes of the underlying assumptions. Furthermore, the expected regime switching time μ_τ and the demand function slope λ are the two parameters that most seriously affect the valuation results. Special care has to be taken with respect to these parameters. However, note that, as for all capital budgeting models, the estimation of input parameters is a permanent source of misvaluation and can lead to wrong decision making.

On the other hand the performed sensitivity analysis also suggests that as more flexibility is considered the results are getting more robust with respect to changes in input parameters. This is an important result on its own. We conjecture that considering flexibility in capital budgeting decisions not only allows management to add value by performing an optimal strategy in reaction to changes of an uncertain environment, but can also provide a certain protection against estimation errors of input parameters. Viewed in that way, real option interaction models can also be a means of reducing modelling risk itself.

6.5   Simulation Results

The question of how the flexibility inherent in a strategic asset can be managed can only be answered in the context of the specific model. For the model of volume flexibility considered here the management of the project is concerned with the optimal capacity expansion and contraction policy. So far we did not explicitly elaborate on the point of how optimal capacity adjustment policies look. Due to the complexity of the different upper and lower thresholds depending on time and currently installed capacity, an analysis of capacity adjustment policies is best illustrated by performing a simulation study. For that reason we trace a typical path of the demand parameter level Θ_t with corresponding regime switching time τ over the whole lifetime of the product life cycle and display the associated optimal operating policies for the four models.

In contrast to the pure evaluation model of project and option values, a simulation study in principle requires the use of stochastic processes for the underlying sources of uncertainty in the *real* world instead of the risk adjusted stochastic

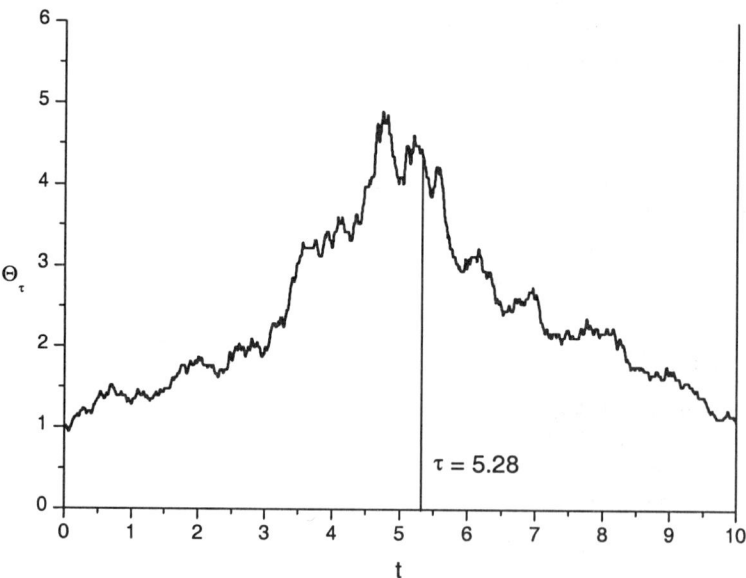

Figure 6.18: Simulated trajectory for the demand parameter level Θ_t of the product life cycle model with regime switching time $\tau = 5.28$ years

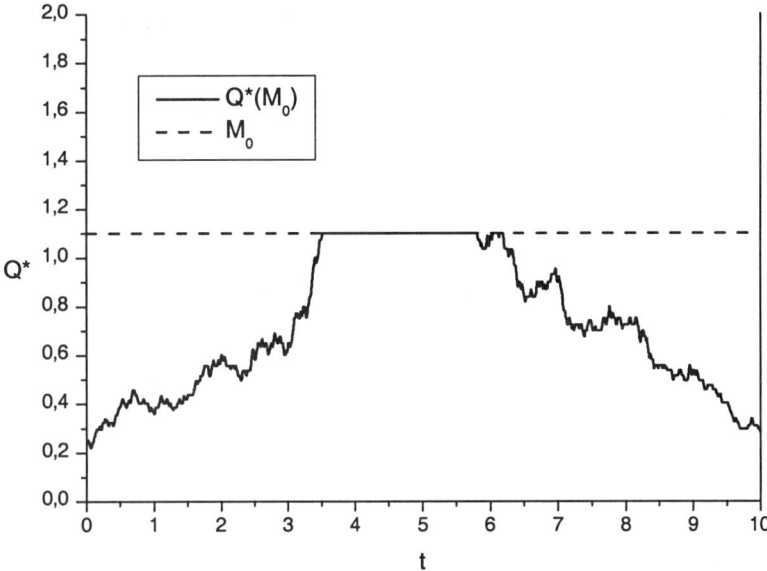

Figure 6.19: Optimal production rate trajectory Q^* for the static project with fixed capacity $M_0 = 1.10$

processes in the *risk neutral* world. This essentially means that we need to know the expected growth rates of the demand parameter level in the growth and decay regime under the original probability measure $\mathbb{P}$, because demand parameter risk is market priced. Since we already assumed risk neutrality in the real world towards the private risk coming from the regime switching uncertainty, no re-adjustment for risk has to be made with respect to regime switching risk. It is interesting to note that the optimal operating policies obtained through maximization of the project value in the risk neutral world are the same as in the original world. Therefore optimization in the risk neutral world not only yields the market price of the project but the optimal operating policy in the *real* world as well. However, there is one serious drawback of simulation studies in the context of real option interactions. Although simulating trajectories of sources of uncertainty in the real world may give us a hint which optimal strategy has to be pursued for a specific realization of the underlying sources of uncertainty, we usually can not use these trajectories to calculate project values because we again face the problem of how to determine the appropriate discount rate in the original world for each operating strategy

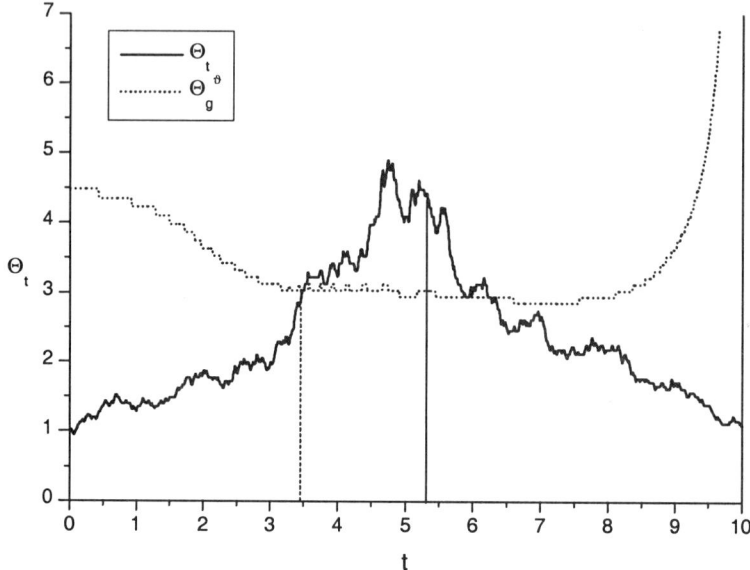

Figure 6.20: Simulated trajectory for the demand parameter level Θ_t of the product life cycle model with timing and intensity option for the static project exercised at time $t = 3.47$ years at demand level $\Theta_g^{\vartheta}(3.47) = 3.034$

chosen.[27] Consequently, if project values need to be calculated the simulation has to be performed in the risk neutral world. On the other hand if we are interested in operating strategies as well as when and how often real options are exercised, simulations have to be performed in the original world.

Since our main focus in studying the optimal operating strategies is in its qualitative nature, we do not introduce the expected growth rates in the original world in order to simulate trajectories but rather to keep the growth rates μ_g and μ_d in the original world for illustrative purposes. The typical trajectory used to compare the optimal policies associated with the four models is displayed in Figure 6.18. The regime switching time for this particular realization is $\tau = 5.28$ years. That is, after 5.28 years the decay regime takes over and the demand parameter Θ_t tends to fall afterwards.

---

[27]This point was neglected by Calistrate/Paulhus/Sick [39] in their simulation study of several simple real options, leading erroneously to deviations from project values in the real and in the risk neutral world.

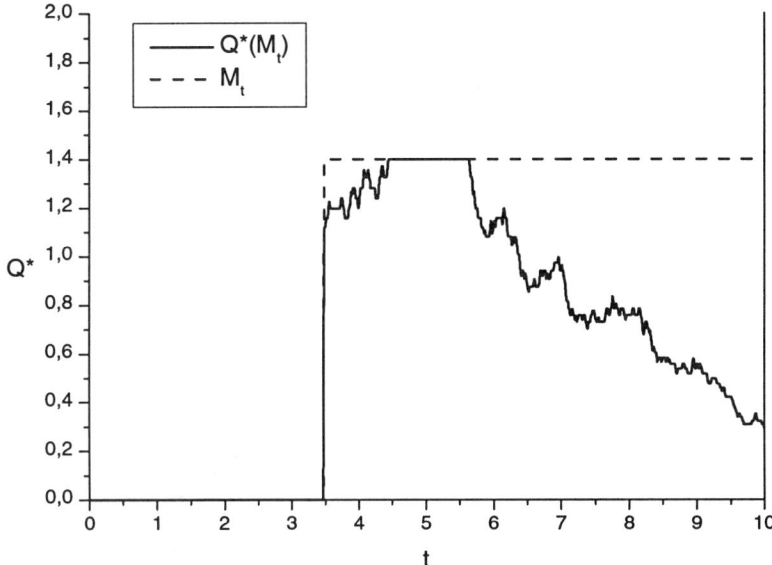

Figure 6.21: Optimal production rate trajectory Q^* for the timing and intensity option of the static project with exercise time $t = 3.47$ and optimal capacity $M_g^\vartheta = 1.40$

First the static project is considered. From above it is known that the optimal initial capacity to install and operate right away is $M_0 = 1.10$. Since the capacity level is kept constant over the whole product life cycle up to terminal time T, this is indicated as the dashed line in Figure 6.19. The corresponding optimal production rate $Q^*(M_0)$ depends, of course, on the realization of the demand parameter Θ_t as well. The basic result is that the firm will be best off by keeping the production rate well below full capacity at times where demand is relatively low, which is the case at the beginning and at the end of the product life cycle. During the middle range of the product life cycle where demand is high, the firm operates at full capacity. The tradeoff between the need to have higher capacity in the middle range and lower capacity at the beginning and the end of the product life cycle becomes evident. For that reason the firm is willing to start with relatively high capacity, although it is not expected to be utilized until the middle range of the product life cycle.

If the timing and intensity option is added to the analysis of the static project the firm can at least reduce the losses during the first years and wait until demand is sufficiently high. For the trajectory displayed in Figure 6.20 the timing and intensity option is exercised at the intersection point between Θ_t and the exercise threshold $\Theta_g^\vartheta(t)$ which happens after 3.47 years. Then the firm starts production with capacity $M_g^\vartheta(t) = 1.40$ which is again represented by the dashed line in Figure 6.21. Since

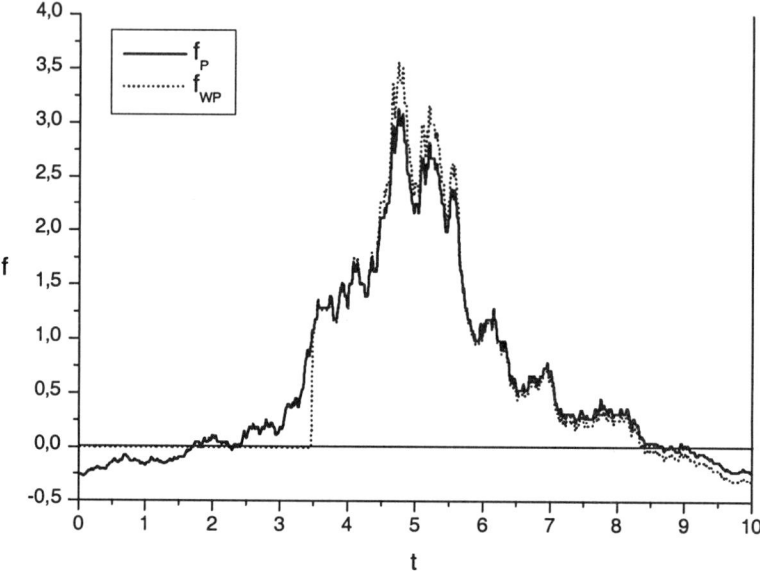

Figure 6.22: Comparison of the cash flow trajectories generated by the static project and the timing and intensity option for the static project

the firm is assumed not to be able to change the capacity afterwards, it remains constant until T. It is obvious how the timing and intensity option changes the strategy the company should follow. By delaying investment the firm can prevent operating inefficiently during the starting phase of the life cycle. In order to reap the high profits in the middle range the intensity of investment is higher than for the static project value. The optimal production rate in the middle of the life cycle is almost near capacity where operation is most efficient. However, there still remains the downside potential at the end of the product life cycle the firm is exposed to. It can only react by downward adjusting production rates.

The protective nature as well as the downside risk of applying a waiting strategy in comparison to the static strategy of investing immediately can be shown by the cash flow profiles generated by both policies in Figure 6.22. First of all, due to waiting the firm prevents earning negative cash flows in the early phase of the product life cycle. Because the firm installs more capacity when it finally exercises the timing and intensity option, it earns larger cash flows in the middle range of the life cycle than with the static operating policy. However, the higher capacity also causes the firm to realize higher losses as the demand finally decreases. Using

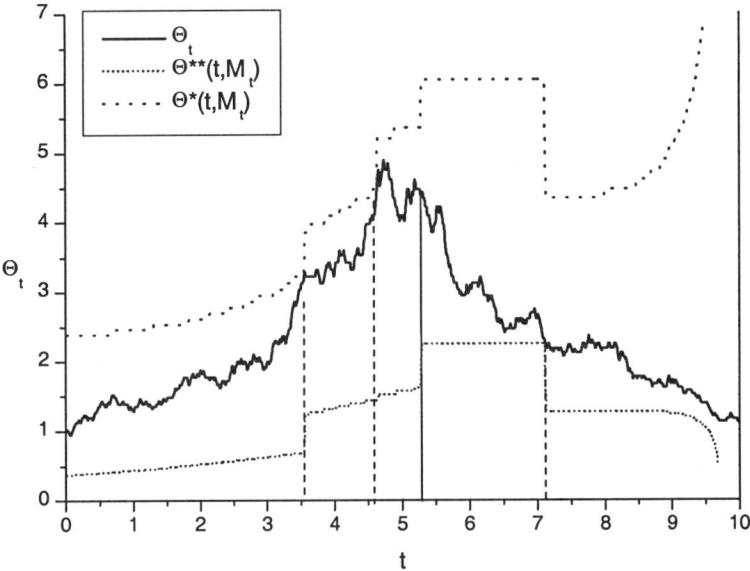

Figure 6.23: Simulated trajectory for the demand parameter level Θ_t of the product life cycle model for the flexible capacity project with capacity expansion at time $t_1 = 3.56$ and $t_2 = 4.64$ and capacity contraction at time $t_3 = 7.12$ and corresponding upper and lower demand parameter thresholds

both resulting cash flow profiles, discounted at the risk free rate r net of discounted investment outlays, results in a time zero value for the static project of $V_g = 2.667$ in comparison to the only slightly higher value of $V_{Wg} = 2.766$ if the timing and intensity option is included. It is worth noting that although the expected value of the model with timing and intensity is significantly larger than the static project value, this need not be the case for all particular realizations of product life cycles. To the contrary, there might frequently occur trajectories where a static strategy would have been better than an operating policy including the timing and intensity option.

Considering next the flexible capacity project, the basic question to answer is which optimal operating policy to apply for the particular sample path of the demand parameter level Θ_t. As can be seen from Figures 6.23 and 6.24 the firm starts operations with an initial capacity of $M_0 = 0.30$. At time $t_1 = 3.47$ the demand parameter in the growth regime reaches the upper threshold level $\Theta_g^*(t, M_t) = \Theta_g^*(3.56, 0.30) = 3.222$ at which capacity expansion to $M_g^*(3.56, 0.30) = 1.00$ is

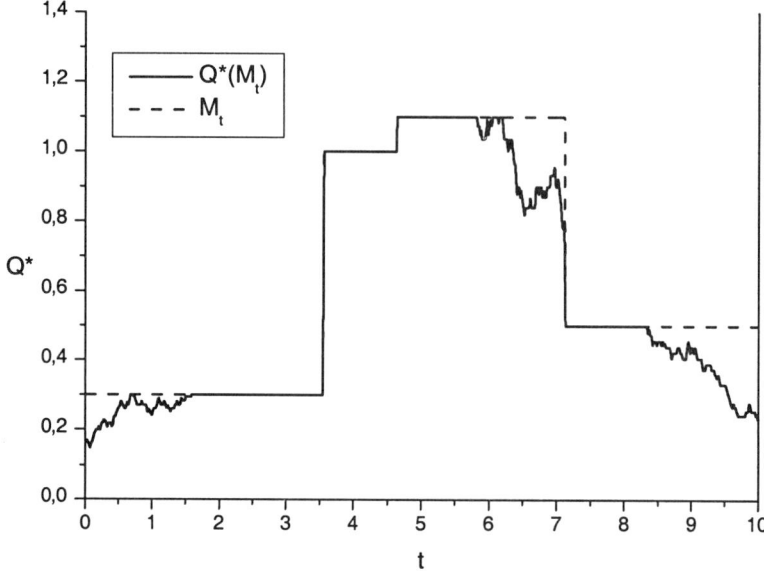

Figure 6.24: Optimal production rate trajectory Q^* for the flexible capacity project with initial capacity $M_0 = 0.30$, capacity expansion at $t_1 = 3.56$ to $M_{Fg}^* = 1.00$ and $t_2 = 4.64$ to $M_{Fg}^* = 1.10$ and capacity contraction at $t_3 = 7.12$ to $M_{Fd}^{**} = 0.50$

optimal. This is displayed by the intersection point where the demand parameter path Θ_t and the upper threshold $\Theta_g^*(t, M_t)$ coincide in Figure 6.23. In terms of impulse control this is the first time when the demand parameter Θ_t reaches the boundary of the continuation region. The corresponding impulse control to apply is to expand capacity to $M_1 = 1.00$ which forces the system back into the continuation region. For $M_1 = 1.00$ the process evolves freely until at time $t_2 = 4.64$ the boundary for the next capacity expansion at demand parameter level $\Theta_g^*(4.64, 1.00) = 4.482$ is hit. The optimal capacity level to switch to is $M_2 = M_g^*(4.64, 1.00) = 1.10$. Until the regime switch occurs at time $\tau = 5.28$, neither the lower boundary nor the upper boundary for capacity adjustment is hit. Since the demand parameter process Θ_t enters the decay regime at time $\tau = 5.28$, the optimal expansion and contraction policies are now different from the growth regime. Thus the upper and lower boundaries of the continuation region change. As expected the firm requires a higher demand level for capacity expansion than in the growth regime. Therefore $\Theta_d^*(5.28, 1.10) = 6.050 > 5.366 = \Theta_g*(5.28, 1, 10)$ and the threshold for capacity expansion admits an upward jump at regime switching time τ. The same is true for the capacity contraction thresholds $\Theta_d^{**}(5.28, 1.10) = 2.248 > 1.616 = \Theta_g^{**}(5.28, 1.10)$ making downward adjustment in capacity more likely than in the growth regime. At time $t_3 = 7.12$ the demand parameter is sufficiently low to hit the contraction boundary $\Theta_d^{**}(7.12, 1.10) = 2.248$ and capacity is adjusted downwards to $M_3 = M_d^{**}(7.12, 1.10) = 0.50$ units. Operating with the reduced capacity of $M_3 = 0.50$, the demand parameter process no longer reaches the boundaries of the continuation region and the firm finally stops production at year $T = 10$ and sells the plant of capacity M_3 for its salvage value.

The optimal production rates corresponding to this optimal operating strategy can be found in Figure 6.24. From there the big advantage of including flexible capacity in an operating strategy becomes obvious. The firm starts at a low capacity level which allows for efficient production at low demand. After only a short time interval of production rates beneath capacity, increasing demand drives production rates to full capacity. A further increase in demand leads the firm to weigh the cost of capacity expansion (investment cost and overhead cost) and the danger of an imminent decrease in production against the foregone profits by operating at low capacity. If demand reaches the point of indifference between both effects, which is exactly the boundary of the continuation region where expansion is triggered, the firm installs additional capacity. For the particular example in the figure the firm increases capacity to $M_1 = 1.00$. Note that the firm still operates at full capacity after the capacity expansion since this is the economically most efficient state of operation. It is more willing to leave some demand unsatisfied rather than set up excess capacity. Although all of our numerical results suggest that capacity changes are always performed in a way such that the firm fully utilizes capacity after adjustments, we are not able to generalize this result. However, for our quite general — and realistic — cost function, this strategy seems to be always optimal. The main condition that causes this phenomenon is the size of the overhead cost parameter c_3. Operating at full capacity means that the fixed cost per unit produced is minimized making higher production rates favorable. As the demand parameter switches to the decay regime the imminent decrease in demand causes the firm to lower production rates in response to a serious decline in prices. As demand further falls, the firm finally contracts its scale of operations by selling capacity to $M_3 = 0.50$. Note that in this case the firm again chooses its new capacity level such that the contracted production facility is fully utilized.

In addition to flexible capacity the timing and intensity option is considered in Figures 6.25 and 6.26. Now the operating strategy of the firm is different. It first waits until the demand parameter process Θ_t hits the investment trigger Θ_{Fg}^{ϑ} which happens after $t_1 = 1.74$ years.[28] The optimal capacity to start production is $M_1 = M_{Fg}^{\vartheta} = 0.60$. At $t_2 = 4.09$ years and at $t_3 = 4.74$ years the demand is high enough to trigger capacity expansion. Once in the decay regime, decreasing

---

[28]Note that the lower threshold for the timing and intensity option does not exist. In Figure 6.25 it was set to zero which is an absorbing barrier that cannot be attained by the process Θ_t.

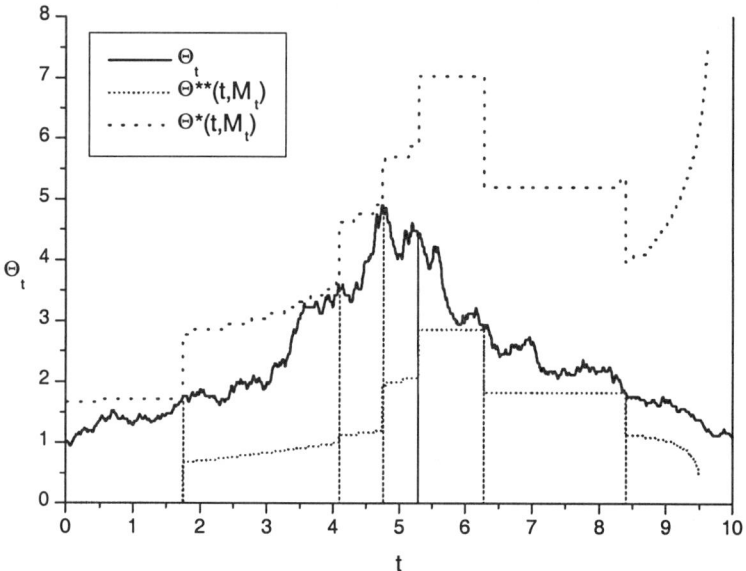

Figure 6.25: Simulated trajectory for the demand parameter level Θ_t of the product life cycle model for the timing and intensity option of the flexible capacity project with exercise of the timing option at $t_1 = 1.74$, capacity expansion at time $t_2 = 4.09$ and $t_3 = 4.74$, and capacity contraction at time $t_4 = 6.28$ and $t_5 = 8.39$ and corresponding upper and lower demand parameter thresholds

demand parameter values result in downwards adjustment of capacity at times $t_4 = 6.26$ and $t_5 = 8.39$. The interpretation is as for the pure flexible capacity project.

Although exactly the same sample path of Θ_t was used to determine the operating strategy for the timing, intensity and flexible capacity project as for the pure flexible capacity project, the operating strategies are different. The reason is that by exercising the timing and intensity option, initial capacity was set to $M_1 = 0.60$ which is different from $M_0 = 0.30$ for the pure flexible capacity project. Once having installed different capacity levels, the subsequent optimal operating strategy differs because being in different operating states means having different expansion and contraction option exercise regions. However, if the timing and intensity was exercised before time $t_1 = 3.56$ such that production started with capacity $M_1 = 0.30$, the subsequent optimal operating strategies would have been the same for both models.

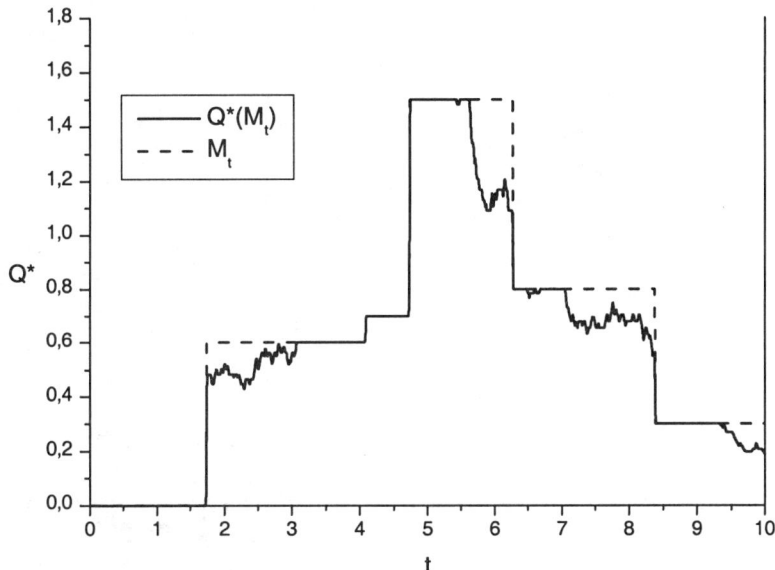

Figure 6.26: Optimal production rate trajectory Q^* for the timing and intensity option of the flexible capacity project with exercise of the timing option at time $t_1 = 1.74$ with initial capacity $M_{Fg}^{\vartheta} = 0.60$, capacity expansion at $t_2 = 4.09$ to $M_{Fg}^* = 0.70$ and $t_3 = 4.74$ to $M_{Fg}^* = 1.50$ and capacity contraction at $t_4 = 6.28$ to $M_{Fd}^{**} = 0.80$ and at $t_5 = 8.39$ to $M_{Fd}^{**} = 0.30$

The optimal production rates as well as the installed capacity levels for the timing, intensity and flexible capacity project are shown in Figure 6.26. Again capacity adjustments — but not the exercise of the intensity option — are always performed to a new capacity level where operating at full capacity is optimal.

Figure 6.27 compares the cash flow profiles of both models resulting from applying the optimal operating strategies. First of all it can be seen that the timing and intensity option helps to prevent negative cash flows at the beginning of the product life cycle. However, this is only a partial advantage because the flexible capacity project already earns positive cash flows long before the timing and intensity option is exercised. Because the capacity installed after the timing and intensity option is exercised is double the capacity already installed in the flexible capacity project the cash flows in the timing and intensity model rapidly increase as demand increases. At time $t_1 = 4.56$ years, capacity is expanded to $M_1 = 1.00$

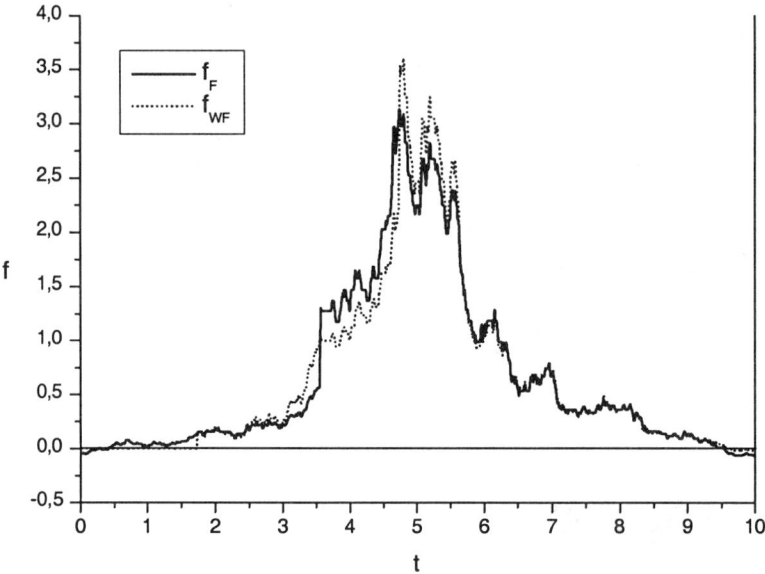

Figure 6.27: Comparison of the cash flow trajectories generated by the flexible capacity project and the timing and intensity option for the flexible capacity project

in the flexible capacity project, which results in larger positive cash flows for the first model. In the middle range of the life cycle the timing and intensity model profits from its higher installed capacity and earns higher cash flows. Finally in the decay regime the cash flow profiles are similar for both models. As cash flows become negative due to low demand, after time 9.5 the timing and intensity model causes fewer losses because it incurs less overhead cost at a capacity of $M_5 = 0.30$ than the pure flexible capacity project with $M_3 = 0.50$.

It has already been discussed that the marginal value introduced by the timing and intensity option in addition to flexible capacity is quite low if possible capacity adjustments are not restricted. This is reflected in the cash flow profiles for both models of this typical product life cycle realization as well. We cannot say from the cash flow profiles directly which strategy would have been favorable. While more flexibility is usually preferable to less flexibility, one would expect that the operating strategy from the timing and intensity model would create more value. However, if the cash flow profiles for this particular example are discounted and adjusted for investment and capacity switching costs, the resulting value for the pure flexible capacity strategy is $V_{Fg} = 3.235$, which is higher than the value of the additional timing and intensity strategy $V_{WFg} = 2.830$. This highlights the fact that more flexibility usually adds *on average* value but not necessarily for each particular realization of the underlying sources of uncertainty.

Another important conclusion can be drawn from comparing the cash flow profiles of the flexible capacity models with the cash flow profiles of the fixed capacity models of Figure 6.23. The fixed capacity models cause significantly higher negative cash flows at the beginning and at the end of the product life cycle than the flexible capacity models. This has several implications for the cash management of the firm. Although we assumed that the firm is all equity financed and can borrow money from the bank unlimited at the risk free interest rate r, this is certainly a simplifying assumption for most real life investment projects. Firms usually consist of a portfolio of different projects that have to be managed in a way such that financing constraints are met. This implies that the firm has to cross-finance new projects at least partially by other ventures generating positive cash flows. For the fixed capacity project much money has to be infused at the beginning and capital is tied up in the project. Since the real options inherent in the project can help to reduce the negative cash flows at the beginning and to generate positive cash flows sooner than with fixed capacity, the firm can use this money to finance other projects. Therefore real options in production accelerate the growth of the firm. This is consistent with findings in the strategy literature. For example the famous BCG-matrix rests on a similar logic.[29] There it is argued that the firm consists of a portfolio of different projects which need to generate the cash flows necessary in order to be able to invest in new high growth ventures. As these so-called *stars* start to generate positive cash flows they turn into *cash cows* which are projects whose positive cash flows can be used to build up new stars. As the success of a cash cow finally fades it turns into a so called *poor dog* which is eliminated from the project portfolio of the firm because it does no longer contribute to financing new stars.[30]. At this point the role of managing real options in production becomes evident. Flexible capacity can help to transform stars sooner into cash cows and keep them there longer which allows the firm to invest in more stars. The general result is that the firm will grow faster and create more value. Viewed in that way real options are of vital strategic importance to the firm.

Since the utilization of production flexibility during a specific realization of the product life cycle need not always be the best operating strategy — an example has just been discussed — calculating expected values for only the four models distinguished seems not sufficient to choose the best operating strategy. Although more flexibility is usually preferable to less flexibility, the incremental value of additional real options is often much less than the cost of acquiring it. For example the timing and intensity option for the static project adds significant value while the timing and intensity option in addition to the flexible capacity project only slightly

---

[29]See e.g., Liebler [144].

[30]The description of the BCG-matrix is not quite complete because for the monopolistic firm considered here we do not take account the market share the firm achieves. However, the analogy is sufficient to demonstrate the main message.

increases expected project values. In the case of competition it has already been argued that this incremental value might completely vanish and a wait-and-see strategy can be outright dangerous. Furthermore, managers may be reluctant to base their decisions on single point estimates associated with a certain real option strategy. In order to choose an operating strategy which dominates other available strategies, simulating the risk profiles of project values is an important tool to analyze the risk management advantages of real options.

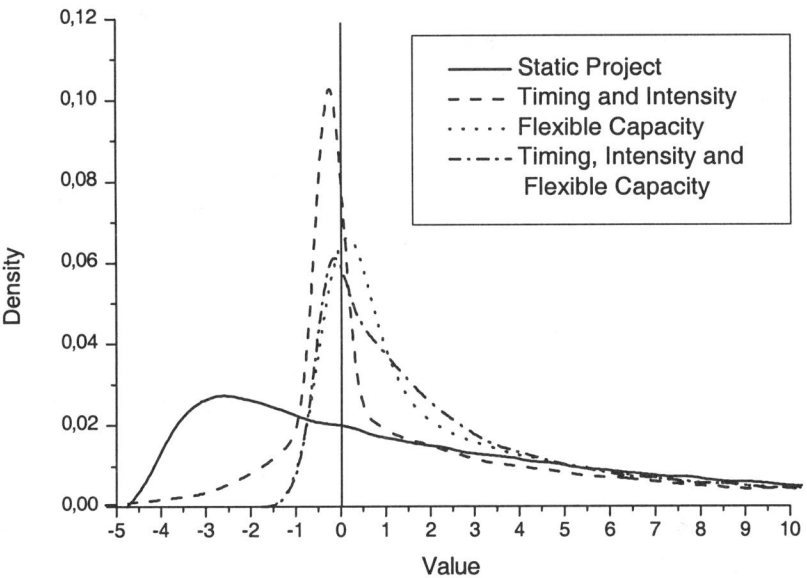

Figure 6.28: Simulated probability density functions (pdf) of the four models with $N = 30000$ computer runs

For the four different operating strategies considered here we use the already determined optimal exercise strategies of the timing, intensity and capacity adjustment options to simulate $N = 30000$ realizations of the stochastic product life cycle and the corresponding values. This results in four risk profiles of project values for each of the above models. The simulated probability density functions (pdf) are displayed in Figure 6.28. Each of the four models admits a risk profile which is skewed to the right. Therefore the upside potential of the project is significant. While all operating strategies have significant probabilities of ending up with a negative project value, the different pdf´s clearly show the downside protection introduced by the different production real options. In comparison to the most risky

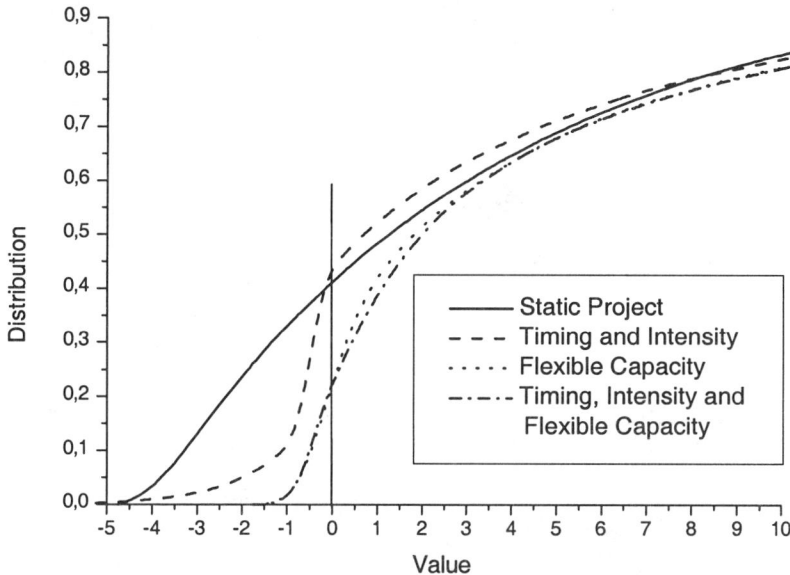

Figure 6.29: Simulated cumulative distribution functions (cdf) of the four models with $N = 30000$ computer runs

strategy, i.e., fixed capacity project, the timing and intensity option seems to be worthwhile to cut very high losses but still yields negative project values with high probability. The figure suggests that the flexible capacity models with and without timing and intensity option are the best operating strategies, which is consistent with our earlier findings.

In fact, from the corresponding cumulative distribution functions displayed in Figure 6.29 it can be directly seen that the probability of a negative project value is approximately 40% in case of the static project, 41% for the timing and intensity option with fixed capacity, 16% for the flexible capacity project and 17% for the timing, intensity and flexible capacity project. However, to know the probabilities of negative project values is only of partial success in finding the optimal strategy. We need to find a means to compare the risk profiles of the four operating strategies. Although risk management tools have been subject to extensive study in the modern finance literature and sophisticated models have been developed, as for example value at risk concepts, coherent risk measures etc.,[31] these methods are of little help

---

[31] See Hull [102], p. 332ff.

in selecting one out of several operating strategies. The problem rather boils down to comparing several distribution functions. These problems are well documented in statistics and we can employ for example *stochastic dominance* criteria to compare the risk profiles of the four models. The following types of stochastic dominance that serve our purposes can be distinguished:[32]

- total dominance,

- first order stochastic dominance and

- second order stochastic dominance.

For the operating strategies considered here total dominance means that a certain operating strategy produces *always*, i.e., for any possible product life cycle realization, higher project values than another operating strategy. Total stochastic dominance is not applicable to our decision problem because it can easily be shown that there exist trajectories where the fixed capacity project yields the highest outcome.

First order stochastic dominance is given if any investor in the risk neutral world prefers one project over another, i.e., let X and Y be two random variables of two value functions, then we say that project X dominates project Y in the first order if and only if $F_X(x) \leq F_Y(x)$ for all x and the inequality is strict for at least one x. We will denote this kind of stochastic dominance by the symbol $>_1$. The intuition behind first order stochastic dominance is quite simple. It says that an operating strategy X is preferable to Y if for each possible outcome x it is more likely that strategy Y is below x than strategy X. First order stochastic dominance can be directly observed from the distribution functions of X and Y if the cdf of X is constantly beneath the cdf of Y. For example in Figure 6.29 the flexible capacity project stochastically dominates the static project in the first order.

If first order stochastic dominance does not apply, a somewhat weaker criterion to compare distributions is second order stochastic dominance. We say that a random variable X stochastically dominates the random variable Y in the second order if and only if

$$\int_{-\infty}^{x_0} (F_Y(x) - F_X(x))\mathrm{d}x \geq 0 \qquad (6.97)$$

for all $x_0 \in \mathbb{R}$ with the inequality being strict for at least one value of x_0. We write $X >_2 Y$. Although second order stochastic dominance does not admit a similar simple interpretation as first order stochastic dominance, it can be shown that in the risk neutral world every investor prefers project X to project Y. In the graph, second order stochastic dominance can be checked by determining the areas between the two distribution functions of X and Y. For example, comparing the static project

---

[32]We will not go too far into the mathematical details of stochastic dominance principles here but rather explain how to work with the machinery. Details of stochastic dominance criterions can be found in Calistrate/Paulhus/Sick [39].

V_P with the timing and intensity project V_W in Figure 6.29 there is a relatively large area between the cdf's of V_P and V_W from -4.5 to -0.5. From -0.5 to 8.0 the cdf of V_W is larger than the cdf of V_P but the area between the curves is not as large as the first area. Therefore the integral in equation (6.96) remains positive. Finally, for values above 8.0 the cdf of V_P becomes larger again and the integral becomes positive. Although these heuristic arguments suggest that the timing and intensity strategy stochastically dominates the fixed capacity strategy in the second order, this is not quite the truth. This is because the timing and intensity project has some slightly higher probability than the fixed capacity project to realize values below -4.5 which can be seen in the figure by the intersection point of the cdf's at -4.5. Therefore the stochastic dominance criterion fails to apply. Nevertheless, we state that the timing and intensity project *almost* stochastically dominates the fixed capacity project in the second order because the probability of observing project values beneath -4.5 is practically negligible.

Finally, if none of the above stochastic dominance criteria is fulfilled we simply call the two distributions *incomparable* which is denoted by <>. Obviously first order stochastic dominance implies second order stochastic dominance but not vice versa. Therefore if it is possible to find a strategy that dominates all other operating strategies in the first or second degree, we will select this strategy as best.

	V_P	V_W	V_F	V_{WF}
V_{WF}	$>_1$ 96.5%	$>_1$ 80.7%	<> 59.6%	
V_F	$>_1$ 95.1%	$>_1$ 80.3%		
V_W	$>_2$ 65.1%			
Theoretical Value	4.057	5.740	7.343	7.427
Simulated Value	4.101	5.746	7.298	7.378

Table 6.4: Comparison of simulation results for base case project values. Cell entries indicate the degree of stochastic dominance of the row entry with respect to the column entry and the proportion of simulations where the row entry yields a higher present value than the column entry.

Table 6.4 shows the stochastic dominance criteria applied to compare the four different operating strategies. The table is constructed such that the column strategy never dominates the row strategy. The percentage values indicate the proportion of simulations in which the row entry yields a higher present value than the column entry. The results confirm our earlier intuition that both flexible capacity projects are preferable to the fixed capacity projects. The timing, intensity and flexible capacity project performs slightly better than the pure flexible capacity project, however, without fulfilling one of the stochastic dominance criteria. We conclude that in the case of unlimited flexible capacity, management is indifferent between applying a pure flexible capacity strategy and a flexible capacity strategy combined with the timing and intensity option.

	Immediate Investment	Timing and Intensity Option
Static Project	$V_{Pg} - c_4 M_0 = 4.057$ $(M_0 = 1.10)$	$V_{Wg} = 5.725$ $(\Theta_g^\vartheta = 2.858, M_g^\vartheta = 3.00)$
Flexible Capacity	$V_{Fg} - c_4 M_0 = 6.089$ $(M_0 = 1.00)$	$V_{WFg} = 6.835$ $(\Theta_{Fg}^\vartheta = 1.994, M_{Fg}^\vartheta = 1.05)$

Table 6.5: Project and option values for capacity restricted to be in the interval $[M_{min}, M_{max}] = [1.00, 3.00]$

	Immediate Investment	Timing and Intensity Option
Static Project	$V_{Pg} - c_4 M_0 = 4.057$ $(M_0 = 1.10)$	$V_{Wg} = 5.725$ $(\Theta_g^\vartheta = 2.858, M_g^\vartheta = 3.00)$
Flexible Capacity	$V_{Fg} - c_4 M_0 = 6.427$ $(M_0 = 1.00)$	$V_{WFg} = 7.034$ $(\Theta_{Fg}^\vartheta = 1.994, M_{Fg}^\vartheta = 1.05)$

Table 6.6: Project and option values for capacity restricted to be in the interval $[M_{min}, M_{max}] = [1.00, 3.00]$ with additional abandonment option

Since unlimited flexibility in capacity adjustments may sometimes be an unrealistic assumption, it is worthwhile to reconsider the flexible capacity models if capacity can only be changed in a certain interval $[M_{min}, M_{max}]$. As a stylized example that shows how the different sources of value change with capacity adjustment restrictions, the interval $[M_{min}, M_{max}] = [1.00, 3.00]$ is chosen. Furthermore, we distinguish the two cases where the firm does and does not possess an additional abandonment option in the flexible capacity case.[33] The resulting expected project values are presented in Tables 6.5 and 6.6. While the static project values do not change in either case because the optimal initial capacity $M_0 = 1.10$ is not affected by the capacity restriction, the timing and intensity option for the static project is slightly lower than in the unrestricted case of Table 6.3 due to the upper limit $M_g^\vartheta = M_{max} = 3.00$. Of course, the largest impact of capacity restriction is on the option value of flexible capacity. For the first case without abandonment option, the project value of the flexible capacity project drops by more than 17% in comparison to the unrestricted case. Although adding an abandonment option recovers some of the lost value the project value with abandonment option is still almost 13% less. As expected the timing, intensity and flexible capacity project is much less sensitive to capacity restrictions because the option to wait at the beginning of the product life cycle prevents management from being forced to enter the market with an inefficiently high initial capacity level. Therefore the loss in value for the first model is only about 8% of the original value. Adding the abandonment option further means that the downside potential at the end of the product life cycle is limited and leads to a decrease in value of only approximately 5%.

	Without Timing	With Timing
Without Abandonment	6.089 (+0%)	6.835 (+12.3%)
With Abandonment	6.427 (+5.5%)	7.034 (+15.5%)

Table 6.7: Project values of the timing and intensity option and the abandonment option added to the flexible capacity project

In comparison to the unrestricted models, capacity restrictions can completely change the perception of optimal operating strategies. The value of the flexible capacity project is dramatically reduced because there are three conditions that destroy value. The first is to be committed to start production immediately at an uneconomically high level of capacity $M_0 = 1.00$ which is caused by the downwards restriction M_{min}. In the course of the product life cycle the firm can

---

[33]An abandonment option can easily be handled by the finite difference software tool by setting the smallest capacity level to zero and the next larger capacity level to $M_{min} = 1.00$.

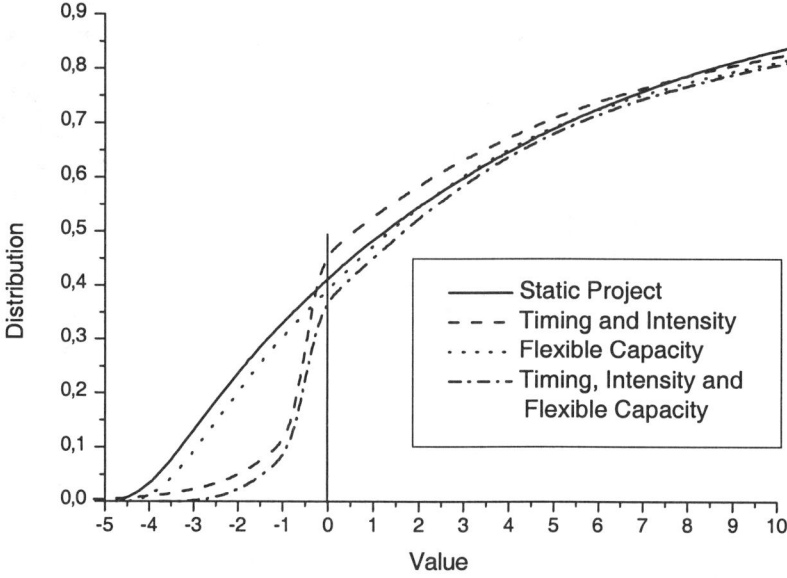

Figure 6.30: Simulated cumulative distribution functions (cdf) of the four models with capacity restriction from $N = 30000$ computer runs

not expand above the upper limit M_{max}, although demand may be high enough to do so. Lastly, at the end of the product life cycle a capacity level beneath M_{min} may again be optimal but cannot be realized. Adding the abandonment option can at least reduce this latter kind of risk. Table 6.7 shows the project values of the different models of flexible capacity and the incremental percentage of value added by the incorporation of the timing and the abandonment option. Unlike the case of unrestricted flexible capacity, the timing and intensity option as well as the abandonment option now add significant value to the overall project. The timing and intensity option is worth more than the abandonment option because it is important at the beginning of the product life cycle, while the abandonment option limits the downside risk of the project at the end of the life cycle. For the symmetric life cycle resulting from the base case parameters, the difference in value of both option results mainly from discounting. As above the option values of the timing and intensity option and the abandonment option are not additive.

With respect to the risk management properties for the restricted flexible capacity models, again a simulation study is performed. The corresponding cumulative probability functions for the case without and the case with abandonment option

	V_P	V_W	V_F	V_{WF}
V_{WF}	$>_1$ 94.3%	$>_1$ 62.6% (20.7%)	$>_1$ 97.0%	
V_F	$>_2$ 88.1%	$<>$ 55.8%		
V_W	$>_2$ 64.6%			
Theoretical Value	4.057	5.725	6.089	6.835
Simulated Value	4.101	5.731	6.060	6.803

Table 6.8: Comparison of simulation results for restricted capacity project values. Cell entries indicate the degree of stochastic dominance of the row entry with respect to the column entry and the proportion of simulations where the row entry yields a higher present value than the column entry. Percentage values in brackets denote the proportion of simulation runs in which column and row entry yield the same outcome.

are displayed in Figures 6.30 and 6.31. In both figures the otherwise identical cdf´s of the static project (solid curve) and the timing and intensity option for the static project (dashed curve) were kept as benchmarks.

Figure 6.30 shows the limited downside protection of flexible capacity (dotted curve) in the presence of capacity adjustment restrictions. The probability of realizing large negative project values is almost as large as for the static project because of the lack of reaction to very low demand levels. Since the option to wait can reduce the startup losses or might even never be exercised if demand develops poorly right from the beginning, it provides a good downside protection. It is easy to see that now the timing, intensity and flexible capacity project represented by the dashed-dotted curve is the best operating strategy because its cdf is beneath all others and therefore stochastically dominates all other strategies in the first order.

Applying the stochastic dominance criterion to the simulated project values yields the results displayed in Table 6.8. In contrast to the unrestricted capacity case the timing, intensity and flexible capacity strategy now clearly dominates all other operating strategies. Although the pure flexible capacity policy performs slightly better than the timing and intensity project with fixed capacity, both strategies

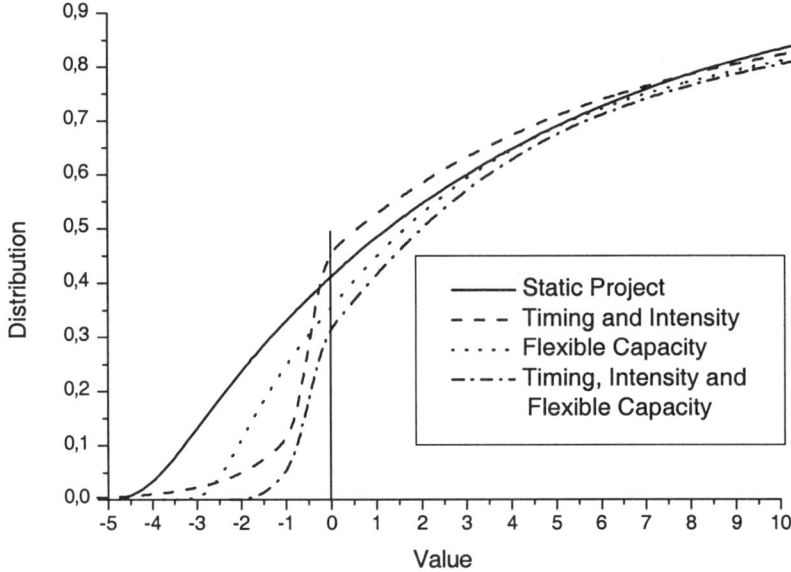

Figure 6.31: Simulated cumulative distribution functions (cdf) of the four models with capacity restriction, additional abandonment option from $N = 30000$ computer runs

are stochastically incomparable. Furthermore, the degree of stochastic dominance of the flexible capacity project with respect to the fixed capacity project is now only of second order. Obviously the imposed capacity restriction seriously affects the performance of flexible capacity. This is apparent from viewing the flexible capacity project as an infinite combination of timing and abandonment options on incremental units of capacity. As the range of possible capacity levels is limited, so is the number of incremental capacity expansion and contraction options and the overall value of the project becomes less.

The impact of adding the abandonment option on both flexible capacity models can be seen from Figure 6.31. The downside protective nature of the abandonment option causes the cdf of the pure flexible capacity project to be shifted to the right. The same, however less pronounced, can be observed for the timing, intensity and flexible capacity project because the incremental value of the abandonment option is now lower. Although the stochastic ordering between the different operating strategies does not change when adding an abandonment option as displayed in Table 6.9, the abandonment option significantly increases the proportions of simu-

	V_P	V_W	V_F	V_{WF}
V_{WF}	$>_1$ 98.1%	$>_1$ 71.8% (16.0%)	$>_1$ 99.3%	
V_F	$>_2$ 91.7%	$<>$ 61.0%		
V_W	$>_2$ 64.6%			
Theoretical Value	4.057	5.723	6.427	7.034
Simulated Value	4.101	5.7312	6.415	7.015

Table 6.9: Comparison of simulation results for restricted capacity project values with additional abandonment option. Cell entries indicate the degree of stochastic dominance of the row entry with respect to the column entry and the proportion of simulations where the row entry yields a higher present value than the column entry. Percentage values in brackets denote the proportion of simulation runs in which column and row entry yield the same outcome.

lations for which the row project yields a higher outcome than the column project. This is obvious for all cases where the row project additionally possesses the abandonment option and the column project not. However, this result is surprising for the comparison between the timing, intensity and flexible capacity project with the pure flexible capacity project. Although the incremental value of the abandonment option is larger for the pure flexible capacity project than for the timing, intensity and flexible capacity project, the number of simulations for which the timing, intensity and flexible capacity project yields a higher outcome increases from 97.0% in Table 6.8 to 99.3% when the abandonment option is added.

The simulation study shows that the rule of thumb that more flexibility usually adds more value applies to the considered production system as well. However, part of an optimal operating strategy in real life situations is not only considering the timing and abandonment of investments as is often done in the real options literature, but also exploiting the value that comes from intensity options and the optimal management of volume flexibility. Although restrictions on the flexibility of capacity may limit the value added by volume flexibility, solely concentrating on timing and abandonment of investments generally leads to neglecting other important sources of value and to misjudgment of optimal operating strategies.

Chapter 7

Conclusions and Extensions

The main purpose of the last chapter was to show how generalized timing and switching options can be used to model real option interactions. We compared four different investment strategies in production facilities of a product whose overall demand is governed by a stochastic product life cycle. Starting from the contingency structure of option interactions, each of the four models was expressed in terms of quasi-variational inequalities. All four models were subsequently implemented numerically using finite difference methods. It was shown how generalized timing and, especially, generalized switching options can be transformed into numerical valuation schemes. Although timing options are often dealt with in the real option literature, solving generalized switching options numerically is a new result on its own. This is because impulse control problems which are behind generalized switching options have not so far been solved numerically using approximations of the quasi-variational inequalities.

Furthermore, in modelling demand as a stochastic product life cycle we treated the subject of dealing with private and market priced risk in real option interaction models. From an economic point of view, product life cycles possess several realistic features that cannot be captured by the usual assumption of demand following a geometric Brownian motion.

In order to analyze the different real option interaction models we performed the following four steps to view the main results and their economic implications:

1. Numerical valuation

2. Sensitivity analysis

3. Strategy space

4. Risk management

The numerical valuation came to the conclusion that the ability to adjust capacity can be of enormous value to management and can not be neglected in favor of certain timing options. The overall result is that more flexibility adds value but that the incremental value of further flexibility is limited.

The performed sensitivity analysis showed two things. First, there are only two input parameters that seriously affect the valuation results. This suggests that special care has to be taken with respect to estimating these parameters. Second and somewhat more surprising, there seems to be evidence that more flexibility not only adds value to the project but also provides a certain protection against misspecification of the model inputs. However, this robustness result is far from being generally valid. It would be worthwhile to pursue this idea further.

The subsequent simulation study enabled us to examine the different operating strategies further. By tracing a sample path of the life cycle forward in time, the differences between the operating strategies became obvious. As expected we found that the ability to expand capacity is particularly important in times of increasing demand, while the capacity contraction option is more important when demand decreases. The special form of the product life cycle emphasizes that, in real life projects, demand often evolves in phases which makes disinvestment options particularly important. However, this would have been neglected in a model where demand is governed by geometric Brownian motion. Another important implication was drawn from the resulting cash flow trajectories for each of the operating strategies. Besides adding value to the project, more flexibility, mainly the flexible capacity option, had very positive effects on the cash management of firms. Managing production flexibility moves cash flows earlier into the positive zone and keeps them there longer. This allows the firm to invest earlier in other ventures and can therefore be a source of superior growth.

Finally, the risk management properties of the four different operating strategies were examined by simulating the probability distribution of the project values. Since expected values alone are not sufficient to judge which operating strategy is favorable, we employed a stochastic dominance criterion to compare the distribution functions of the four models. Because capacity adjustments were assumed to be unrestricted, both flexible capacity projects performed significantly better than the projects with fixed capacity. However, the timing and intensity option added to the flexible capacity project could practically not be distinguished from the pure flexible capacity project. The situation changed as we restricted capacity to be in a certain range. Then the timing, intensity and flexible capacity project dominated all others.

Besides its illustrating function as a real option interaction model, the presented study of production flexibility can also serve as a building block for more realistic models of operating flexibility. From a technical point of view it would be worthwhile to consider effects of competition and time lags on capacity adjustments. Furthermore, the model can be extended to more products by introducing process flexibility or to take account of international trade. This would also include the study of a portfolio of projects (interproject dependencies) which have to be managed in parallel or sequentially (growth options).[1] Both extensions would introduce more sources of uncertainty whose numerical treatment is subject to further

---

[1] See Childs [51] or Childs/Ott/Triantis [50] for the valuation of interrelated projects.

research. Moreover, the model of production should also incorporate flexibilities that are situated behind the production stage in the value chain of the firm. For example, keeping inventories may help to smooth production rates and reduces the need to adjust capacity instantaneously.[2]

Another topic that can be vitally integrated into the model are marketing efforts. The firm may for example have the ability to reduce the private regime switching risk by further investing in market research. This would introduce a learning option where the firm needs to invest in order to resolve private risk. Marketing expenditures can also be seen as a means to influence the product life cycle endogenously. Placing ads may, e.g., change the shape of the product life cycle directly. The optimal stopping and impulse control models introduced in Section 3.2 are capable of this kind of endogeneity, because marketing efforts can be modelled as actions with a direct impact on the underlying sources of uncertainty. However, from a capital market point of view the determination of an equivalent martingale measure places some serious problems in the direct impact case.[3] Finally, the impact of taxes, depreciation and financing decisions on the optimal capacity adjustment policies could be worthwhile to investigate. During the study we assumed that the decision making firm is all equity financed. However, relaxing this assumption can lead to two important extensions. First, the flexibility to choose the optimal capital structure dynamically adds value to the firm by utilizing the tax shield of debt financing and may subsequently alter the optimal operating policy.[4] Second, dropping the assumption of an all equity financed firm can lead to agency problems because operating strategies that maximize firm value usually do not coincide with strategies that maximize the value for the equityholders.[5]

All of the above mentioned extensions can be modelled using the stochastic control framework and its refinements proposed in this book. More importantly these findings are not restricted to optimal production strategies but can be applied to all other real option interaction models as well. Using the graphical representation of real option interactions together with the mathematical formulation of generalized timing and generalized switching options allows us to easily solve complex option interaction models numerically or analytically, provided the capital market link is appropriately performed. The stochastic control framework has several advantages over other approaches.

From a theoretical point of view it admits the necessary mathematical rigor

---

[2]This point was first made by Pennings [184], p. 89. He deals with the trade-off between production flexibility and inventories.

[3]We did not explicitly elaborate on this kind of resulting market incompleteness in the book; see e.g. Hubalek/Schachermayer [99] for a criticism of the complete markets assumption. From a strategic point of view it should be emphasized that a real option theory in incomplete markets is crucial for extending the analysis of real options with a reactive nature to the proactive management of real options with a strategic focus. That means, the next generation of real option models should include the endogeneity of the interplay of a firm's actions and its environment rather than solely concentrating on the reactions of the firm in response to changes in the exogenously given environment.

[4]See Mauer/Triantis [157].

[5]See Mauer/Ott [156].

to place its results on solid grounds. In resting on the theory of optimal stopping and impulse control, the whole machinery of stochastic control can be applied. Furthermore, it turned out that large parts of modern financial theory and especially real options are nothing else but stochastic control problems merged with the condition of no arbitrage in capital markets.

From a practical point of view one of the main implications of the stochastic control framework is that decision situations involving real options can be decomposed into generalized timing and switching options. Arbitrary real option models can be composed using these two building blocks. For the real option approach to achieve a broader acceptance among practitioners it is crucial to develop models that are able to cope with the complexity of real life business decisions. The proposed stochastic control framework possesses the ability to do so and in distinguishing only two main modelling elements, the method is especially suitable for computational implementation. Furthermore, the graphical decomposition method can be used to develop a graphical user interface that facilitates the implementation and communication of complex real option interactions.

In summary, the stochastic control framework for valuing real options is capable of systematizing the treatment of real option interactions occurring in arbitrary applications. It therefore represents an extension of the previous work on real option pricing from a theoretical as well as from a practical point of view. To fully exploit its potential in theory and application is the subject of future research.

Bibliography

[1] Andrew B. Abel. A stochastic model of investment, marginal Q, and the market value of the firm. *International Economic Review*, 25(2):305–323, 1985.

[2] Andrew B. Abel, Avinash K. Dixit, Janice C. Eberly, and Robert S. Pindyck. Options, the value of capital, and investment. *Quarterly Journal of Economics*, 64:753–777, 1996.

[3] Andrew B. Abel and Janice C. Eberly. An exact solution for the investment and value of a firm facing uncertainty, adjustment costs and irreversibility. *Journal of Economic Dynamics and Control*, 21(4+5):831–852, 1997.

[4] Luis H. R. Alvarez. Demand uncertainty and the value of supply opportunities. *Journal of Economics*, 64(2):163–175, 1996.

[5] Luis H. R. Alvarez. Optimal exit and valuation under demand uncertainty: A real options approach. *European Journal of Operational Research*, 114:320–329, 1999.

[6] Martha Amram and Nalin Kulatilaka. Disciplined decisions: Aligning strategy with the financial markets. *Harvard Business Review*, 77(1):95–104, 1999.

[7] Martha Amram and Nalin Kulatilaka. *Real Options: Managing Strategic Investment in an Uncertain World*. Harvard Business School Press, Boston, Massachusetts, 1999.

[8] Martha Amram and Nalin Kulatilaka. Uncertainty: The new rules for strategy. *Journal of Business Strategy*, (3):25–29, 1999.

[9] Fridrik M. Baldursson. Irreversible investment under uncertainty in oligopoly. *Journal of Economic Dynamics and Control*, 22:627–644, 1998.

[10] Carliss Y. Baldwin. Optimal sequential investment when capital is not readily reversible. *Journal of Finance*, 37(3):763–782, 1982.

[11] Carliss Y. Baldwin, Scott P. Mason, and Richard S. Ruback. Evaluation of government subsidies to large scale energy projects: A contingent claims approach. Technical Report 66, Harvard Business School, 1983.

[12] Avner Bar-Ilan and William C. Strange. Investment lags. *The American Economic Review*, 86(3):610–622, 1996.

[13] Avner Bar-Ilan and William C. Strange. Urban development with lags. *Journal of Urban Economics*, 39:87–113, 1996.

[14] Avner Bar-Ilan and William C. Strange. A model of sequential investment. *Journal of Economic Dynamics and Control*, 22:437–463, 1998.

[15] Avner Bar-Ilan and William C. Strange. The timing and intensity of investment. *Journal of Macroeconomics*, 21(1):57–77, 1999.

[16] Avner Bar-Ilan and Agnes Sulem. Explicit solution of inventory problems with delivery lags. *Mathematics of Operations Research*, 20(3):709–720, 1995.

[17] Eric D. Beinhocker. Robust adaptive strategies. *Sloan Management Review*, (2):95–106, 1999.

[18] Giovanni Beliossi and Han Smit. Using real options: From project evaluation to security analysis. Talk given at Unicom conference "Real Options and Investment Decisions under Uncertainty", London (1999).

[19] Gregory K. Bell. Volatile exchange rates and the multinational firm: Entry, exit, and capacity options. In Lenos Trigeorgis, editor, *Real Options in Capital Investment - Models, Strategies, and Applications*, pages 163–183. Praeger, Westport, Connecticut, 1995.

[20] Alain Bensoussan and Jacques-Louis Lions. *Impulse Control and Quasi-Variational Inequalities*. Bordas, Paris, 1984.

[21] Alain Bensoussan and Charles S. Tapiero. Impulsive control in management: Prospects and applications. *Journal of Optimization Theory and Applications*, 37(4):419–442, 1982.

[22] Philip G. Berger, Eli Ofek, and Itzhak Swary. Investor valuation of the abandonment option. *Journal of Financial Economics*, 42(2):257–287, 1996.

[23] Jonathan Berk, Richard C. Green, and Vasant Naik. Valuation and return dynamics of R&D ventures. School of Business Administration, University of Washington, 1997.

[24] Jonathan B. Berk, Richard C. Green, and Vasant Naik. Optimal investment, growth options, and security returns. *The Journal of Finance*, 54(5):1553–1607, 1999.

[25] Antonio Bernardo and Bhagwan Chowdhry. Resources, real options and corporate strategy. Technical Report 33-98, The Anderson School at UCLA, 1999.

[26] Nick H. Bingham and Rüdiger Kiesel. *Risk-Neutral Valuation*. Springer-Verlag, London, 1998.

[27] Fischer Black and Myron Scholes. The pricing of options an corporate liabilities. *Journal of Political Economy*, 81(3):637–654, 1973.

[28] Nicolas P. B. Bollen. Real options and product life cycles. To appear in *Management Science*, 1998.

[29] Nicolas P. B. Bollen. Valuing options in regime-switching models. *The Journal of Derivatives*, (Fall 1998):38–49, 1998.

[30] Svetlana I. Boyarchenko and Sergei Z. Levendorskiĭ. Entry and exit strategies under non-Gaussian distributions. In Michael J. Brennan and Lenos Trigeorgis, editors, *Project Flexibility, Agency, and Competition*, pages 71–84. Oxford University Press, New York, Oxford, 2000.

[31] Richard A. Brealey and Stewart C. Myers. *Principles of Corporate Finance*. McGraw-Hill, New York, 1996.

[32] Kjell Arne Brekke and Bernt Øksendal. The high contact principle as a sufficiency condition for optimal stopping. In Lund Diderik and Bernt Øksendal, editors, *Stochastic Models and Option Values*, pages 187–208. Elsevier Science Publishers, Amsterdam, The Netherlands, 1991.

[33] Michael J. Brennan and Eduardo S. Schwartz. Finite difference methods and jump processes arising in the pricing of contingent claims: A synthesis. *Journal of Financial and Quantitative Analysis*, 13:461–474, 1978.

[34] Michael J. Brennan and Eduardo S. Schwartz. Evaluating natural resource investments. *Journal of Business*, 58(2):135–157, 1985.

[35] Michael J. Brennan and Lenos Trigeorgis(Ed.). *Project Flexibility, Agency, and Competition: New Developments in the Theory and Application of Real Options*. Oxford University Press, New York, Oxford, 2000.

[36] Adrian Buckley. *International Investment - Value Creation and Appraisal: A Real Options Approach*. Copenhagen Business School Press, Copenhagen, 1998.

[37] J. S. Busby and C. G. C. Pitts. *Assessing Flexibility in Capital Investment*. The Chartered Institute of Management Accountants Publishing, London, 1998.

[38] Ricardo J. Caballero and Robert S. Pindyck. Uncertainty, investment, and industry evolution. *International Economic Review*, 37(3):641–662, 1996.

[39] Dan Calistrate, Marc Paulhus, and Gordon Sick. Real options for managing risk: Using simulation to characterize gain in value. Technical report, Real Options Group, 1999.

[40] Dan Calistrate, Marc Paulhus, and Gordon Sick. A recombining binomial tree for valuing real options with complex structures. Technical report, Real Options Group, 1999.

[41] Jeanette Capel. How to service a foreign market under uncertainty: A real option approach. *European Journal of Political Economy*, 8:455–475, 1992.

[42] Dennis Capozza and Yuming Li. The intensity and timing of investment: The case of land. *American Economic Review*, 84(4):889–904, 1994.

[43] Peter Carr. The valuation of sequential exchange opportunities. *Journal of Finance*, 43(5):1235–1256, 1988.

[44] Peter Carr. The valuation of American exchange options with application to real options. In Lenos Trigeorgis, editor, *Real Options in Capital Investment - Models, Strategies, and Applications*, pages 109–120. Praeger, Westport, Connecticut, 1995.

[45] Peter Carr and Dimitri Faguet. Valuing finite-lived options as perpetual. Technical report, Cornell University, Ithaca, NY, 1996.

[46] Jorge A. Chan-Lau and Peter B. Clark. Fixed investment and capital flows: A real options approach. Technical Report WP/98/125, International Monetary Fund, 1998.

[47] Andrew H. Chen, John W. Kensinger, and James A. Conover. Valuing flexible manufacturing facilities as options. *The Quarterly Review of Economics and Finance*, 38(Special Issue):651–674, 1998.

[48] Joseph A. Cherian, Jayendu Patel, and Ilya Khripko. Optimal extraction of nonrenewable resources when costs cumulate. In Michael J. Brennan and Lenos Trigeorgis, editors, *Project Flexibility, Agency, and Competition*, pages 224–253. Oxford University Press, New York, Oxford, 2000.

[49] Paul D. Childs, Steven H. Ott, and Timothy J. Riddiough. Valuation and information acquisition policy for claims written on noisy real assets. Technical report, University of Kentucky, Lexington, Kentucky, 1999.

[50] Paul D. Childs, Steven H. Ott, and Alexander J. Triantis. Capital budgeting for interrelated projects: A real options approach. *Journal of Financial and Quantitative Analysis*, 33(3):305–334, 1998.

[51] Paul David Childs. *Capital Budgeting for Interrelated Projects in a Real Options Framework*. PhD thesis, University of Wisconsin-Madison, 1995.

[52] Kee H. Chung and Charlie Charoenwong. Investment options, assets in place, and the risk of stocks. *Financial Management*, 20(1):21–33, 1991.

[53] Kee H. Chung and Kyu H. Kim. Growth opportunities and investment decisions: A new perspective on the cost of capital. *Journal of Business Finance & Accounting*, 24(3+4):413–424, 1997.

[54] Morris A. Cohen and Arnd Huchzermeier. Global supply chain management: A survey of research and applications. In S. Tayur, M. Magazine, and R. Ganeshan, editors, *Quantitative Models for Supply Chain Management*, chapter 21, pages 789–821. Kluwer Academic Press, 1998.

[55] Morris A. Cohen and Arnd Huchzermeier. Global supply chain network management under price/exchange rate risk and demand uncertainty. Technical report, Otto-Beisheim Graduate School of Management, WHU Koblemz, 1999.

[56] George M. Constantinides. Stochastic cash management with fixed and proportional transaction costs. *Management Science*, 22(12):1320–1331, 1976.

[57] George M. Constantinides and Scott F. Richard. Existence of optimal simple policies for discounted-cost inventory and cash management in continuous time. *Operations Research*, 26(4):620–636, 1978.

[58] Thomas E. Copeland and Philip T. Keenan. How much is flexibility worth? *The McKinsey Quarterly*, (2):38–49, 1998.

[59] Thomas E. Copeland and Philip T. Keenan. Making real options real. *The McKinsey Quarterly*, (3):128–141, 1998.

[60] Tom Copeland, Tim Koller, and Jack Murrin. *Valuation: Measuring and Managing the Value of Companies*. John Wiley & Sons, New York, 1990.

[61] Gonzalo Cortazar and Eduardo Schwartz. Monte Carlo evaluation model of an undeveloped oil field. Technical Report 1, The Anderson School at UCLA, 1998.

[62] Gonzalo Cortazar, Eduardo Schwartz, and Andrés Löwener. Optimal investment and production decisions and the value of the firm. Technical Report 2, The Anderson School at UCLA, 1997.

[63] Gonzalo Cortazar and Eduardo S. Schwartz. A compound option model of production and intermediate inventories. *Journal of Business*, 66(4):517–540, 1993.

[64] Gonzalo Cortazar, Eduardo S. Schwartz, and Marcelo Salinas. Evaluating environmental investments: A real options approach. *Management Science*, 44:1059–1070, 1998.

[65] Hugh Courtney, Jane Kirkland, and Patrick Viguerie. Strategy under uncertainty. *Harvard Business Review*, (11 + 12):67–79, 1997.

[66] Graham A. Davis. Estimating volatility and dividend yield when valuing real options to invest or abandon. *The Quarterly Review of Economics and Finance*, 38(Special Issue):725–754, 1998.

[67] M. H. A. Davis and A. R. Norman. Portfolio selection with transaction costs. *Mathematics of Operations Research*, 15(4):676–713, 1990.

[68] Marco A. G. Dias and Katia M. C. Rocha. Petroleum concession with extendible option: Investment timing and value using mean reversion and jump processes for oil prices. Technical report, Petrobas, Petroleo Brasileiro S.A., 1998.

[69] Avinash Dixit. Entry and exit decisions under uncertainty. *Journal of Political Economy*, 97(3):620–638, 1989.

[70] Avinash Dixit. Intersectoral capital allocation under price uncertainty. *Journal of International Economics*, 23:309–325, 1989.

[71] Avinash Dixit. Investment and hysteresis. *Journal of Economic Perspectives*, 6(1):107–132, 1992.

[72] Avinash Dixit and Robert S. Pindyck. *Investment under Uncertainty*. Princeton Universitiy Press, Princeton, New Jersey, 1994.

[73] Avinash Dixit and Robert S. Pindyck. The options approach to capital investment. *Harvard Business Review*, (5 + 6):105–115, 1995.

[74] Avinash Dixit, Robert S. Pindyck, and Sigbjørn Sødal. A markup interpretation of optimal rules for irreversible investment. Technical Report 5971, National Bureau of Economic Research, 1997.

[75] Darrell Duffie. *Dynamic Asset Pricing Theory*. Princeton University Press, Princeton, New Jersey, 1996.

[76] Jerome F. Eastham and Kevin J. Hastings. Optimal impulse control of portfolios. *Mathematics of Operations Research*, 13(4):588–605, 1988.

[77] Y.H. Farzin, K.J.M. Huisman, and P.M. Kort. Optimal timing of technology adoption. *Journal of Economic Dynamics and Control*, 22:779–799, 1998.

[78] Kay M. Fischer. *Realoptionen: Anwendungsmöglichkeiten der Finanziellen Optionstheorie auf Realinvestitionen im In- und Ausland*. PhD thesis, Hamburg, 1996. Diss.

[79] Nicolai J. Foss. The resource-based perspective: An assessment and diagnosis of problems. DRUID Working Paper No. 97-1, Copenhagen Business School, 1997.

[80] Robert Geske. The valuation of corporate liabilities as compound options. *Journal of Financial and Quantitative Analysis*, 12:541–552, 1977.

[81] Robert Geske. The valuation of compound options. *Journal of Financial Economics*, 7(1):63–81, 1979.

[82] I. V Girsanov. On transforming a certain class of stochastic processes by absolutely continuous substitution of measures. *Theory of Probability Applied*, 5:285–301, 1960.

[83] Steven R. Grenadier. The strategic exercise of options: Development cascades and overbuilding in real estate markets. *The Journal of Finance*, 51(5):1653–1679, 1996.

[84] Steven R. Grenadier. Equilibrium with time-to-build. In Michael J. Brennan and Lenos Trigeorgis, editors, *Project Flexibility, Agency, and Competition*, pages 275–296. Oxford University Press, New York, Oxford, 2000.

[85] Steven R. Grenadier and Allen M. Weiss. Investment in technological innovations: An option pricing approach. *Journal of Financial Economics*, 44(3):397–416, 1997.

[86] Christian Großmann and Hans-Görg Roos. *Numerik partieller Differentialgleichungen*. Teubner Verlag, Stuttgart, Germany, 1992.

[87] J. Michael Harrison. *Brownian Motion and Stochastic Flow Systems*. John Wiley & Sons, New York, 1985.

[88] J. Michael Harrison and D. M. Kreps. Martingales and arbitrage in multiperiod securities markets. *Journal of Economic Theory*, 20:381–408, 1079.

[89] J. Michael Harrison and S. R. Pliska. Martingales and stochastic integrals in the theory of continuous trading. *Stochastic Processes and Applications*, 11:215–260, 1981.

[90] J. Michael Harrison and S. R. Pliska. A stochstic calculus model of continuous trading: Complete markets. *Stochastic Processes and Applications*, 15:313–316, 1983.

[91] Michael J. Harrison and Jan A. Van Mieghem. Multi-resource investment strategies: Operational hedging under demand uncertainty. *European Journal of Operational Research*, 113:17–29, 1999.

[92] Michael J. Harrison, Thomas M. Sellke, and Allison J. Taylor. Impulse control of Brownian motion. *Mathematics of Operations Research*, 8(3):454–466, 1983.

[93] Hua He and Robert S. Pindyck. Investments in flexible production capacity. *Journal of Economic Dynamics and Control*, 16:575–599, 1992.

[94] Jimmy E. Hilliard and Jorge Reis. Valuation of commodity futures and options under stochastic convenience yields, interest rates, and jump diffusions in the spot. *Journal of Financial and Quantitative Analysis*, 33(1):61–86, 1998.

[95] David G. Hobson. Stochastic volatility. Technical report, School of Mathematical Sciences, University of Bath, 1996.

[96] James E. Hodder and Alexander Triantis. Valuing flexibility: An impulse control framework. *Annals of Operations Research*, 45:109–130, 1993.

[97] Andreas Höger. The effect of interest rate uncertainty and taxation on the optimal timing of investment. Technical report, Vienna University of Economics and Business Administration, Austria, 1998.

[98] Yaozhong Hu and Bernt Øksendal. Optimal time to invest when the price processes are geometric Brownian motion. *Finance and Stochastics*, 2(3):295–310, 1998.

[99] F. Hubalek and Walter Schachermayer. The limitations of no-arbitrage arguments for real options. Technical Report 95, University of Vienna, Austria, 1999.

[100] Arnd Huchzermeier and Morris A. Cohen. Valuing operational flexibility under exchange rate risk. *Operations Research*, 44(1):100–113, 1996.

[101] Arnd Huchzermeier and Christoph H. Loch. Evaluating R&D projects as learning options: Why more variability is not always better. Technical report, Otto-Beisheim Graduate School of Management, WHU Koblenz, Germany, and INSEAD, Fontainebleau, France, 1999.

[102] John C. Hull. *Options, Futures, and Other Derivatives*. Prentice-Hall, Upper Saddle River, New Jersey, 1997.

[103] Simon R. Hurst, Eckhard Platen, and Svetlozar T. Rachev. Option pricing for a logstable asset price model. Technical Report, University of California Santa Barbara, 1995.

[104] Jonathan Ingersoll and Stephen A. Ross. Waiting to invest: Investment and uncertainty. *Journal of Business*, 65(1):1–29, 1992.

[105] Monique Jeanblanc-Picqué. Impulse control method and exchange rate. *Mathematical Finance*, 2(3):161–177, 1993.

[106] Bardia Kamrad. A lattice claims model for capital budgeting. *IEEE Transactions on Engineering Management*, 42(2):140–148, 1995.

[107] Bardia Kamrad and Ricardo Ernst. Multiproduct manufacturing with stochastic input prices and output yield uncertainty. In Lenos Trigeorgis, editor, *Real Options in Capital Investment - Models, Strategies, and Applications*, pages 281–302. Praeger, Westport, Connecticut, 1995.

[108] Bardia Kamrad and Shreevardhan Lele. Production, operating risk and market uncertainty: A valuation perspective on controlled policies. *IIE Transactions*, 30:455–468, 1998.

[109] Ioannis Karatzas and Steven E. Shreve. *Methods of Mathematical Finance*. Springer-Verlag, New York, Berlin, Heidelberg, 1998.

[110] W. Carl Kester. Today s options for tomorrow s growth. *Harvard Business Review*, (3 + 4):153–160, 1984.

[111] W. Carl Kester. Turning growth options into real assets. In Raj Aggarwal, editor, *Capital Budgeting under Uncertainty*, chapter 11, pages 187–207. Prentice Hall, Englewood Cliffs, New Jersey, 1993.

[112] Michael Kilka. *Realoptionen - Optionspreistheoretische Ansätze bei Investitionsentscheidungen unter Unsicherheit*. Schriftenreihe der SGZ-Bank. Fritz Knapp Verlag, Frankfurt a. M., 1995.

[113] Myung-Jig Kim, Young-Ho Oh, and Robert Brooks. Are jumps in stock returns deversifiable? Evidence and implications for option pricing. *Journal of Financial and Quantitative Analysis*, 29:609–631, 1994.

[114] Thomas S. Knudsen, Bernhard Meister, and Mihail Zervos. On the realtionship of the dynamic programming approach and the contingent claim approach to asset valuation. *Finance and Stochastics*, 3(4):433–449, 1999.

[115] Christian Koch. *Optionsbasierte Unternehmensbewertung: Realoptionen im Rahmen von Akquisitionen*. Gabler, Wiesbaden, 1999.

[116] Bruce Kogut. Joint ventures and the option value to expand and acquire. *Management Science*, 37:19–33, 1991.

[117] Bruce Kogut and Nalin Kulatilaka. Operating flexibility, global manufacturing, and the option value of a multinational network. *Management Science*, 40(1):123–139, 1994.

[118] Bruce Kogut and Nalin Kulatilaka. Options thinking and platform investments: Investing in opportunity. *California Management Review*, (4):52–71, 1994.

[119] Bruce Kogut and Nalin Kulatilaka. Capabilities as real options. Technical report, Wharton School, University of Pennsylvania and School of Management, Boston University, 1997.

[120] Ralf Korn. Optimal impulse control when control actions have random consequences. *Mathematics of Operations Research*, 22(3):639–667, 1997.

[121] Ralf Korn. Portfolio optimization with strictly positive transaction costs and impulse control. *Finance and Stochastics*, 2:85–114, 1998.

[122] Peter M. Kort. Optimal R&D investments of the firm. *OR Spektrum*, 20:155–164, 1998.

[123] Nalin Kulatilaka. Valuing the flexibility of flexible manufacturing systems. *IEEE Transactions on Engineering Management*, 35(4):250–257, 1988.

[124] Nalin Kulatilaka. The value of flexibility: The case of a dual-fuel industrial steam boiler. *Financial Management*, 22(3):271–280, 1993.

[125] Nalin Kulatilaka. Operating flexibilities in capital budgeting: Substitutability and complementarity in real options. In Lenos Trigeorgis, editor, *Real Options in Capital Investment - Models, Strategies, and Applications*, pages 121–132. Praeger, Westport, Connecticut, 1995.

[126] Nalin Kulatilaka. The value of flexibility: A general model for real options. In Lenos Trigeorgis, editor, *Real Options in Capital Investment - Models, Strategies, and Applications*, pages 89–108. Praeger, Westport, Connecticut, 1995.

[127] Nalin Kulatilaka and A. J. Marcus. General formulation of corporate real options. *Research in Finance*, pages 183–199, 1988.

[128] Nalin Kulatilaka and Alan J. Marcus. Project valuation under uncertainty: When does DCF fail? *Journal of Applied Corporate Finance*, 5(3):92–100, 1992.

[129] Nalin Kulatilaka and Stephen Gary Marks. The strategic value of flexibility: Reducing the ability to compromise. *American Economic Review*, 78(3):574–580, 1988.

[130] Nalin Kulatilaka and Enrico C. Perotti. Strategic growth options. *Management Science*, 44:1021–1031, 1998.

[131] Nalin Kulatilaka and Enrico C. Perotti. Time-to-market capability as a Stackelberg growth option. Technical report, School of Management, Boston University and University of Amsterdam, 1999.

[132] Nalin Kulatilaka and Lenos Trigeorgis. The general flexibility to switch: Real options revisited. *The International Journal of Finance*, 6:778–798, 1994.

[133] Nalin Kulatilaka and N. Venkatraman. Are you preparing to compete in the New Economy? Use a real options navigator. Technical report, Systems Research Center, Boston University School of Management, 1998.

[134] Damien Lamberton and Bernard Lapeyre. *Introduction to Stochastic Calculus Applied to Finance*. Chapman & Hall, London, New York, 1996.

[135] Bart Lambrecht and William Perraudin. Real options and preemption under incomplete information. Technical report, University of Cambridge, JIMS and Birbeck College, London, 1999.

[136] Bart M. Lambrecht. The impact of debt financing on entry and exit in a duopoly. Technical report, University of Cambridge, JIMS, 1998.

[137] Bart M. Lambrecht. Strategic sequential investments and sleeping patents. Technical report, University of Cambridge, JIMS, 1998.

[138] Bart M. Lambrecht and William Perraudin. Real options and preemption. Technical report, University of Cambridge, JIMS and Birbeck College, London, 1997.

[139] Diane M. Lander and George E. Pinches. Challenges to the practical implementation of modeling and valuing real options. *The Quarterly Review of Economics and Finance*, 38(Special Issue):537–567, 1998.

[140] Richard S. M. Lau. Strategic flexibility: A new reality for world-class manufacturing. *SAM Advanced Management Journal*, 2:11–15, 1996.

[141] John V. Leahy. Investment in competitive equilibrium: The optimality of myopic behavior. *The Quarterly Journal of Economics*, 108(4):1105–1133, 1993.

[142] Ya-Kang Lawrence Lee. *Extensions of Option Pricing Theory to the Analysis of New Business Opportunities*. PhD thesis, Stanford University, 1997.

[143] Keith J. Leslie and Max P. Michaels. The real power of real options. *The McKinsey Quarterly*, pages 4–22, 1997.

[144] Hans Liebler. *Strategische Optionen: Eine kapitalmarktorientierte Bewertung von Investitionen unter Unsicherheit*. Universitätsverlag Konstanz, Konstanz, 1996.

[145] J. Lintner. The valuation of risky assets and the selection of risky investments
       in stock portfolios and capital budgets. *Review of Economics and Statistics*,
       47:13–37, 1965.

[146] John Liu and Dongqing Yao. Strategic options in re-engineering of a man-
       ufacturing system with uncertain completion time. *European Journal of
       Operational Research*, 115:47–58, 1999.

[147] Timothy A. Luehrman. Investment opportunities as real options: Getting
       started on the numbers. *Harvard Business Review*, 76(4):51–67, 1998.

[148] Timothy A. Luehrman. Strategy as a portfolio of real options. *Harvard
       Business Review*, 76:89–99, 1998.

[149] Morten W. Lund. Real options in offshore oil field development projects.
       Technical report, Natural Gas Marketing & Supply, Statoil; Stavanger, Nor-
       way, 1999.

[150] Saman Majd and Robert S. Pindyck. Time to build, option value, and in-
       vestment decisions. *Journal of Financial Economics*, 18:7–27, 1987.

[151] Saman Majd and Robert S. Pindyck. The learning curve and optimal pro-
       duction under uncertainty. *Rand Journal of Economics*, 20(3):331–343,
       1989.

[152] Benoit Mandelbrot and H. M. Taylor. On the distribution of stock price
       differences. *Operations Research*, 15:1057–1062, 1967.

[153] W. Margrabe. The value of an option to exchange one asset for another.
       Journal of Finance, 33(1):177–186, 1978.

[154] Scott P. Mason and Carliss Y. Baldwin. Evaluation of government subsidies
       to large-scale energy projects: A contingent claims approach. *Advances in
       Futures and Operations Research*, 3:169–181, 1988.

[155] Michael J. Mauboussin. Get real: Using real options in security analysis.
       Technical Report 10, Credit Suisse First Boston Corporation, 1999.

[156] David C. Mauer and Steven H. Ott. Agency costs, underinvestment, and
       the optimal capital structure. In Michael J. Brennan and Lenos Trigeorgis,
       editors, *Project Flexibility, Agency, and Competition: New Developments
       in the Theory and Application of Real Options*, pages 151–179. Oxford
       University Press, New York, Oxford, 2000.

[157] David C. Mauer and Alexander J. Triantis. Interactions of corporate fi-
       nancing and investment decisions: A dynamic framework. *The Journal of
       Finance*, 49(4):1253–1277, 1994.

[158] Robert McDonald and Daniel Siegel. Option pricing when the underlying asset earns a below-equilibrium rate of return: A note. *The Journal of Finance*, 39(1):261–265, 1984.

[159] Robert McDonald and Daniel Siegel. Investment and the valuation of firms when there is an option to shut down. *International Economic Review*, 26(2):331–349, 1985.

[160] Robert McDonald and Daniel Siegel. The value of waiting to invest. *Quarterly Journal of Economics*, pages 707–727, 1986.

[161] Robert L. McDonald. Real options and rules of thumb in capital budgeting. In Michael J. Brennan and Lenos Trigeorgis, editors, *Project Flexibility, Agency, and Competition*, pages 13–33. Oxford University Press, New York, Oxford, 1998.

[162] Robyn McLaughlin and Robert A. Taggart. The opportunity cost of using excess capacity. *Financial Management*, 21(2):12–23, 1992.

[163] Florian Meise. *Realoptionen als Investitionskalkül*. R. Oldenbourg Verlag, München, Wien, 1998.

[164] Antonio S. Mello, John E. Parsons, and Alexander J. Triantis. An integrated model of multinational flexibility and financial hedging. *Journal of International Economics*, 39(1):27–51, 1995.

[165] Robert C. Merton. Theory of rational option pricing. *Bell Journal of Economics and Management Science*, 4(1):141–183, 1973.

[166] Robert C. Merton. *Continuous Time Finance*. Basil Blackwell, Cambridge, Oxford, 1990.

[167] Gilbert E. Metcalf and Kevin A. Hassett. Investment under alternative return assumptions: Comparing random walks and mean reversion. *Journal of Economic Dynamics and Control*, 19:1471–1488, 1995.

[168] R. Metters. Producing multiple products with stochastic seasonal demand and capacity limits. *Journal of the Operational Research Society*, 49(3):263–272, 1998.

[169] Kristian R. Miltersen and Eduardo S. Schwartz. Pricing of options on commodity futures with stochastic term structure of convenience yields and interest rates. *Journal of Financial and Quantitative Analysis*, 33(1):33–59, 1998.

[170] Stefan Mittnik and Svetlozar T. Rachev, editors. *Stable Models in Finance*, New York, 1999. Rodin.

[171] Stefan Mittnik, Svetlozar T. Rachev, and Marc Paolella. Stable Paretian modeling in finance: Some empirical and theoretical aspects. In Robert J. Adler, Raisa E. Feldman, and Murad S. Taqqu, editors, *A Practical Guide to Heavy Tails*, pages 79–110. Birkhäuser, Basel, 1998.

[172] Alberto Moel and Peter Tufano. Bidding for the antamina mine: Valuation and incentives in a real options context. Technical report, Graduate School of Business Administration, Harvard University, 1998.

[173] Ernesto Mordecki. Optimal stopping for a diffusion with jumps. *Finance and Stochastics*, 3:227–236, 1999.

[174] J. Mossin. Equilibrium in a capital asset market. *Econometrica*, 35:768–783, 1966.

[175] Gabriela Mundaca and Bernt Øksendal. Optimal stochastic control with application to the exchange rate. *Journal of Mathematical Economics*, 29(2):225–243, 1998.

[176] Stewart C. Myers. Determinants of corporate borrowing. *Journal of Financial Economics*, 5:147–175, 1977.

[177] Stewart C. Myers. Finance theory and financial strategy. *Interfaces*, 14(1):126–137, 1984.

[178] Stewart C. Myers and Saman Majd. Abandonment value and project life. *Advances in Futures and Operations Research*, 4:1–21, 1990.

[179] Nancy A. Nichols. Scientific management at Merck: An interview with CFO Judy Lewent. *Harvard Business Review*, 72(1):88–99, 1994.

[180] Bernt Øksendal. *Stochastic Differential Equations: An Introduction with Applications*. Springer-Verlag, New York, Berlin, Heidelberg, 1992.

[181] Richard E. Ottoo. Valuation of internal growth opportunities: The case of a biotechnology company. *The Quarterly Review of Economics and Finance*, 38(Special Issue):615–633, 1998.

[182] James L. Paddock, Daniel R. Siegel, and James L. Smith. Option valuation of claims on real assets: The case of offshire petroleum leases. *Quarterly Journal of Economics*, pages 479–508, 1988.

[183] Ariel Pakes. Patents as options: Some estimates of the value of holding European patent stocks. *Econometrica*, 54(4):755–784, 1986.

[184] Enrico Pennings. *Real Options and Managerial Decision Making*. PhD thesis, Rotterdam Institute of Business Economic Studies, Erasmus University of Rotterdam, 1998.

[185] Enrico Pennings and Onno Lint. The option value of advanced R&D. *European Journal of Operational Research*, 103:83–94, 1997.

[186] Robert S. Pindyck. Irreversible investment, capacity choice, and the value of the firm. *American Economic Review*, 78(5):969–985, 1988.

[187] Robert S. Pindyck. Investments of uncertain costs. *Journal of Financial Economics*, 34:53–76, 1993.

[188] Robert S. Pindyck. Inventories and the short-run dynamics of commodity prices. *RAND Journal of Economics*, 25(1):141–159, 1994.

[189] Laura Quigg. Optimal land development. In Lenos Trigeorgis, editor, *Real Options in Capital Investment - Models, Strategies, and Applications*, pages 265–280. Praeger, Westport, Conecticut, 1995.

[190] Ariane Reiss. An option pricing approach to investments in innovations in a competitive environment. Technical report, Universität Tübingen, 1997.

[191] Peter Ritchken and Bardia Kamrad. A binomial contingent claims model for valuing risky ventures. *European Journal of Operational Research*, 53:106–118, 1991.

[192] Joseph A. Ritter and Joseph G. Haubrich. Commitment as investment under uncertainty. Working Paper 9606, Federal Reserve Bank of Cleveland, 1996.

[193] Stephen A. Ross. Uses, abuses, and alternatives to the net-present-value rule. *Financial Management*, 24(3):96–102, 1995.

[194] Bryan R. Routledge, Duane J. Seppi, and Chester S. Spatt. Equilibrium forward curves for commodities. Graduate School of Industrial Administration, Carnegie Mellon University, 1998.

[195] Jacob Sagi. The interaction between quality control and production. Technical report, University of British Columbia, Vancouver, Canada, 1998.

[196] Ronald A. Sanchez. *Strategic Flexibility, Real Options, and Product-Based Strategy*. PhD thesis, MIT, 1991.

[197] Ronald A. Sanchez. Strategic flexibility in product competition. *Strategic Management Journal*, 16:135–159, 1995.

[198] Eduardo S. Schwartz. The stochastic behavior of commoditiy prices: Implications for valuation and hedging. *The Journal of Finance*, 52(3):923–973, 1997.

[199] Eduardo S. Schwartz. Valuing long-term commodity assets. *Financial Management*, 27(1):57–66, 1998.

[200] A. K. Sethi and S. P. Sethi. Flexibility in manufacturing: A survey. *The International Journal of Flexible Manufacturing Systems*, 2:289–328, 1990.

[201] David J. Sharp. Uncovering hidden value in high-risk investments. *Sloan Management Review*, (2):69–74, 1991.

[202] W. Sharpe. Capital asset prices: A theory of market equilibrium under conditions of risk. *Journal of Finance*, 19:425–442, 1964.

[203] David C. Shimko. *Finance in Continuous Time: A Primer*. Kolb Publishing Company, Miami, Florida, 1992.

[204] Steven Shreve, Prasad Chalasani, and Somesh Jha. Stochastic calculus in finance. Draft, Carnegie Mellon University, 1997.

[205] Gordon Sick. Real options. In R. Jarrow et al., editor, *Finance*, chapter 21, pages 631–691. Elsevier Science Publishers B. V., Amsterdam, New York, Oxford, 1995.

[206] Gordon Sick. Analyzing a real option on a petroleum property. Technical report, University of Calgary, Canada, 1999.

[207] Han T. J. Smit. Investment analysis of offshore concessions in The Netherlands. *Financial Management*, 26(2):5–17, 1997.

[208] Han T. J. Smit and L. A. Ankum. A real options and game-theoretic approach to corporate investment strategy under competition. *Financial Management*, 22(3):241–250, 1993.

[209] James E. Smith and Kevin F. McCardle. Valuing oil properties: Integrating option pricing and decision analysis approaches. *Operations Research*, 46(2):198–217, 1998.

[210] James E. Smith and Kevin F. McCardle. Options in the real world: Lessons learned in evaluating oil and gas investments. *Operations Research*, 47(1):1–15, 1999.

[211] James E. Smith and Robert F. Nau. Valuing risky projects: Option pricing theory and decision analysis. *Management Science*, 41(5):795–816, 1995.

[212] Kenneth Smith and Alexander Triantis. Untapped options for creating value in acquisitions. *Mergers and Acquisitions*, 29(3):17–20, 1994.

[213] Sigbjørn Sødal. A simplified exposition of smooth pasting. *Economics Letters*, 58:217–223, 1998.

[214] Agnes Sulem. Explicit solution of a two-dimensional deterministic inventory problem. *Mathematics of Operations Research*, 11(1):134–146, 1986.

[215] Agnes Sulem. A solvable one-dimensional model of a diffusion inventory system. *Mathematics of Operations Research*, 11(1):125–133, 1986.

[216] Elizabeth Olmsted Teisberg. Capital investment strategies under uncertain regulation. *Rand Journal of Economics*, 24(4):591–604, 1993.

[217] Sheridan Titman. Urban land prices under uncertainty. *American Economic Review*, 75(3):505–514, 1985.

[218] O. Tourinho. The option value of reserves of natural resources. Technical Report Working Paper 94, University of California, Berkley, 1979.

[219] Alexander J. Triantis and James E. Hodder. Valuing flexibility as a complex option. *Journal of Finance*, 45(2):549–565, 1990.

[220] Lenos Trigeorgis. *Valuing Real Investment Opportunities: An Options Approach to Strategic Capital Budgeting*. PhD thesis, Graduate School of Business Administration, Harvard University, 1986.

[221] Lenos Trigeorgis. Anticipated competitive entry and early preemptive investment in deferrable projects. *Journal of Economics and Business*, 43:143–156, 1991.

[222] Lenos Trigeorgis. A log-transformed binomial numerical analysis method for valuing comlex multi-option investments. *Journal of Financial and Quantitative Analysis*, 26:309–326, 1991.

[223] Lenos Trigeorgis. The nature of option interactions and the valuation of investments with multiple real options. *Journal of Financial and Quantitative Analysis*, 28(1):1–20, 1993.

[224] Lenos Trigeorgis. Real options and interactions with financial flexibility. *Financial Management*, 22(3):202–224, 1993.

[225] Lenos Trigeorgis. *Managerial Flexibility and Strategy in Resource Allocation*. The MIT Press, Cambridge, Massachusetts; London, England, 1996.

[226] Lenos Trigeorgis and Scott P. Mason. Valuing managerial flexibility. *Midland Corporate Finance Journal*, 5(1):14–21, 1987.

[227] Lenos Trigeorgis (Ed.). *Real Options in Capital Investment - Models, Strategies, and Applications*. Praeger, Westport, Connecticut, 1995.

[228] Jan A. VanMieghem. Investment strategies for flexible resources. *Management Science*, 44:1071–1078, 1998.

[229] Alexander Vollert. On Margrabe's option to exchange in a Paretian-stable subordinated market. *Mathematical and Computer Modelling*, 34:1185–1197, 2001.

[230] Helen Weeds. Reverse hysteresis: R&D investment with stochastic innovation. Technical report, Fitzwilliam College, University of Cambridge, UK, 1999.

[231] Helen Weeds. Strategic delay in a real options model of R&D competition. Technical report, Fitzwilliam College, University of Cambridge, UK, 1999.

[232] Peter J. Williamson. Strategy as options on the future. *Sloan Management Review*, (2):117–126, 1999.

[233] Ram Willner. Valuing start-up venture growth options. In Lenos Trigeorgis, editor, *Real Options in Capital Investment - Models, Strategies, and Applications*, pages 221–241. Praeger, Westport, Conecticut, 1995.

[234] Ramzi Zein. *Investment Timing under Uncertainty: Real Options, Process Specification and Convenience Yields*. PhD thesis, Graduate School, Yale University, 1998.

Index